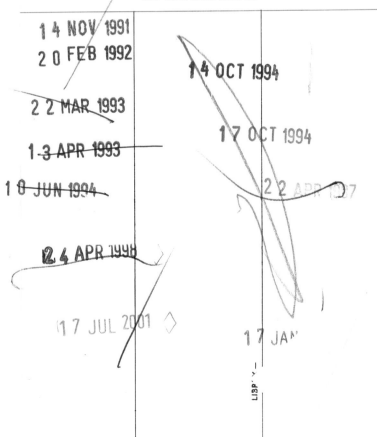

ENVIRONMENTAL MONITORING:
MEETING THE TECHNICAL CHALLENGE

ENVIRONMENTAL MONITORING: MEETING THE TECHNICAL CHALLENGE

Editor: E M Cashell

Proceedings of the International Conference organised by
ISA International (Ireland Section)

8 May 1990
Cork, Ireland

Supported by the Institute of Physics (Irish Branch and ISAT Group)
and Cork Regional Technical College

IOP Short Meetings Series No 29
Institute of Physics

Copyright © 1990 by IOP Publishing Ltd and individual contributors unless otherwise stated. All rights reserved. Multiple copying of the contents or parts thereof without permission is in breach of copyright but permission is hereby given to copy titles and abstracts of papers and names of authors. Permission is usually given upon written application to IOP Publishing Ltd to copy illustrations and short extracts from the text of individual contributions, provided that the source (and, where appropriate, the copyright) is acknowledged.

CODEN — IPS SE3

British Library Cataloguing in Publication Data

CIP catalogue record for this book is available from the British Library

ISBN 0-85498-529-8
ISSN 0269-8986

Published by IOP Publishing Ltd
Techno House, Redcliffe Way, Bristol BS1 6NX, United Kingdom

Printed in the United Kingdom by Dotesios (Printers) Ltd

Contents

Preface

Part 1: Trends in sensors and instrumentation

1   Recent developments in new sensors for environmental monitoring
    R Briggs

55 ISFET–based chemical sensors for environmental monitoring
    J A Voorthuyzen

67 In-situ continuous fibre optic sensors for water pollutants
    B MacCraith

81 Fibre optic sensors for environmental monitoring: Can they meet the task?
    K T V Grattan

Part 2: The challenge for industry

107 Monitoring beyond the boundary fence
    B Callan

137 Environmental management and loss measurement in dairy processing
    J Palmer

145 Air quality monitoring in the UK
    J Wilken

147 The legal and policy framework for environmental monitoring in the European Community
    L Cashman

157 NETT: a key for information on environmental technology
M C Nas

Preface

This volume presents the proceedings of the international conference on Environmental Monitoring which was held in Cork, Ireland, May 8th., 1990. Environmental issues are now an area of major interaction and indeed conflict between the public and modern industry. Increasingly, the onus is being placed on science to find technological solutions to environmental problems: more stringent legislation is being introduced to satisfy public demands, forcing industry and the monitoring authorities to seek better ways for assessing air and water quality.

This conference examined the current state of the technology and methodology employed by industry to fulfill its legal and public responsibilities and looked at new, innovative technologies which are likely to figure prominently in environmental monitoring in the future. Public demands for a better quality of life and therefore a cleaner environment have led to new and more stringent legislation, with the European Commission, in particular, playing a key role here. Since the Single European Act, environmental protection policy has become an integral part of all other European Community policies; monitoring provides the technological link between the legal consequences of this policy and industrial reality. There are two papers in the volume which discuss the nature and development of this legislation (Cashman) and, conversely, the market opportunity which has been generated by the enhanced need for environmental protection (Nas).

It is likely that sensors with greater sensitivity and specificity will be required in the future and that on-line monitoring will become necessary if the legal demands are to be fully met. Areas of technology which are showing particular signs of promise in this regard are fibre-optic based sensors and solid state devices: the conference was fortunate in being able to invite prominent research scientists in the field who contributed four papers which examine the current state of the technology and the likelihood of reliable and workable sensors becoming available which can satisfy the legal requirements. In a general review, Briggs looked at current trends in environmental monitoring for the water industry in particular. Recent developments in the application of solid state devices such as isfets were described by Voorthuyzen. A general review of fibre-optic based sensors was presented by Grattan and some specific applications were examined by McCraith.

Industries with a strong need for reliable and sensitive detection of pollutants include the chemical and food processing industries, both of which play a prominent role in the economic life of the Cork region. In his contribution, Callan discussed the technology and methodology employed in one chemical company and in his contribution, Palmer discussed the critical role of waste management and control within the food processing industry.

The meeting was organised by ISA International (Ireland Section), in collaboration with the Institute of Physics (Instrument Science and Technology Group), the Institute of Physics (Irish Branch) and Cork Regional Technical College. The meeting was held in conjunction with the annual exhibition of instrumentation for measurement and control organised by ISA International (Ireland Section) and was directed at technical and management personnel from industry, as well as academic interests. As a consequence, it is gratifying to note that over half the 108 delegates were from industry, with a particularly strong representation from the chemical industry.

The organisation of an international conference requires the support and assistance of many people, particularly the members of the Conference Committee. Special thanks is therefore extended to G. Dullea, B. Goggin, K. Grattan, L. McDonnell. T. O'Doherty, A. Petersen, H. Makin and S. Linehan. Finally, I would like to thank Ms. Mary Harney, Minister for Environmental Protection, who opened the conference and welcomed the delegates.

E.M. Cashell

RECENT DEVELOPMENTS IN NEW SENSORS FOR ENVIRONMENTAL MONITORING

By Dr R Briggs

Professor and Head of Water and

Environmental Instrumentation Research

Measurement and Instrumentation Centre

City University London

ABSTRACT

Water and water using Industries' requirements for monitoring instrumentation and control systems are discussed and gaps in available technology highlighted.

Recent developments in sensors and support systems are examined in the context of the perceived gaps in available measurement and control technology needed for domestic and trade waste treatment and for the monitoring and control of environmental pollution generally.

Particular attention is given to improved sampling and sample preparation techniques of relevance and also to methods of rendering sensors more robust and reliable, including automatic cleaning and calibration techniques. The role of microprocessors and personal computers is examined also and the possibility of transferring technologies of relevance from other fields such as medicine and aviation explored.

Recent developments in electrochemistry and electro-optics are examined and their applicability to air and water pollution monitoring and control discussed. So too are biologically based sensors and systems including the use of bacteria, invertebrates and fish. The possibility of adopting certain immuno-assay techniques to the measurement of trace organics and other toxicants is explored. Finally the use of technologies utilising aircraft and satellite mounted active laser systems in air and water pollution monitoring is examined.

INTRODUCTION

In recent years, emphasis has been placed by the UK Water Industry on the need for increased use of instrumentation in the operation of many of its process plants. This desire to promote efficiency and become more cost effective has increased significantly with the privatisation of the Industry in 1989. Not only has the need for more precise process control and the remote supervision of unmanned plants increased, but, in addition, with the setting up by Government of the National Rivers Authority, so too has the need for reliable instrumentation for monitoring source and receiving waters including both permanent station, portable and, in some cases, submersible equipment.

Additionally, with the recently increased powers and capabilities of the various Inspectorates and Health and Safety organisations, a clear need has arisen for equipment for monitoring ambient air conditions and head space above bodies of water and in partially filled conduits such as sewers. Although this paper is concerned primarily with water pollution aspects, attention will be drawn to gaseous measurements of relevance, remembering the interrelationship between air and water pollution that can occur.

In most applications, and in water industry applications in particular, the performance of ICA equipment as a whole, especially that of the front end sensors has been regarded as far from satisfactory, the faults being attributable in part to the technical quality of the equipment supplied and also to type of personnel expected to use and maintain it. Additionally, it has become clear that there is an absence of relevant user specifications which could be taken on board by the Instrument Industry and this also has contributed significantly to the overall dissatisfaction expressed.

As a result, following the implementation of the recommendations made in the Report of the DOE/NWC Working Party on Control Systems for the Water Industry (1), the Water Industry as a whole and the Water Research Centre (WRC) in particular, has and is still expending considerable effort and monies on the production of the required specifications and on the setting up and utilising of two major evaluation and demonstration facilities (EDFS). Details of progress to date in the above areas has been published (2) and in addition, much effort has been directed by the WRC and elsewhere (3) to the identification of water Industry and Water Users' needs and to studying recent developments in relevant sensor and system technology. A review of these developments in the context of satisfying identified needs follows:

WATER INDUSTRY AND WATER USER'S NEEDS

The requirement for water quality data has been assessed with consideration of the following topics:

(i)   Proof of compliance with relevant legislation

(ii) Measurements needed for the operation and maintenance of water and waste water treatment and transport facilities

(iii) Control of industrial effluent treatment

(a)   <u>Sampling Requirements</u>

These are detailed in the joint circular from the Department of the Environment and the Welsh Office (1982). This document draws attention to the EC Directive relating to the quality of water intended for Human Consumption (80/778/EEC) the text of which is reproduced in the document and also gives guidance on the way in which the directive is to be implemented in England and Wales. The document contains a list of the required determinands under the following headings:

- Organoleptic
- Physio-chemical
- Toxic Substances
- Microbiological

Proof of compliance is achieved in most cases by means of a programme of sampling and subsequent laboratory analysis and there appears to be ample instrumentation available to carry out reference methods of analysis at the present time. However, most Water Authority Scientists are of the opinion that far lower limits of detection will be required in the future, particularly for monitoring toxic substances. There are certain determinands that are better measured on site including microbiological variables and labile determinands such as dissolved oxygen, nitrate ions and ammonia. Additionally, there is a need in some cases for continuous monitoring, for intake protection for example, and this implies a need for field-based staff unless the required instrumentation and telemetry equipment are available in a robust and reliable form.

If portable battery powered versions of the required instruments could readily be plugged into existing and future telemetry networks (e.g. PTSN, private wire, UHF radio or satellite) then the results of field analysis and field calibration could be automatically loaded into a central data file with obvious savings in time and man hours. This also would appear to be a promising area for the application of new sensor technologies.

(b) <u>Operational Needs</u>

Most Water Authorities, public Utilities and other relevant government agencies are committed to a policy of continued improvement to the environment and this implies better process control and in consequence an increased need for reliable process control instrumentation.

The matters to be considered in this context are:

(i) The likelihood of the existence, in developing countries, of increasing surpluses of unskilled labour, permanent shortages of capital for public works and ever increasing costs.

(ii) The ability to treat wastewater to the minimum standard required to maintain the quality of the receiving waters downstream of the discharge at a level adequate for the use to which the water is to be put. Clearly this will be different for different rivers and different sections of river.

Comprehensive details of measurement requirements in both water and wastewater treatment areas have been published recently. This information is based on reports received from Canadian, West German, Swedish and United Kingdom representatives and the contained data have been combined in the form of a listing of variable of most significance in treatment process control, measurement priorities and equipment availability (4, 5, 6).

A list of variables of most significance and ranked according to priority has recently been produced by the Process Instrumentation Specification and Evaluation Group (PISEG) and the ICA Policy Group of the Association of Water Authorities and their Liaison Engineers have been utilised by WRC in the writing of user specifications. However, because of the impending impact of EC legislation, this list has been extended to include heavy metals and trace organics thought to be potentially harmful to Health (3) and is provided in tabular form below (Table 1).

TABLE 1: Variables of Greatest Significance in the Operation of Water and Waste Water Treatment and Transport Systems.

Water Treatment and Distribution	Sewage Treatment	Sludge Disposal	Source and Receiving Waters
Chlorine	Dissolved Oxygen		
Ozone			
Organic Matter	Suspended Solids	Heavy Metals	Suspended Solids
pH Value	Sludge Solids		Nitrate ion
Colour	Organic Matter		Trace Organiser
Turbidity	Ammonia	Pathogens	pH value
Aluminium	Nitrate ion		Algae
Iron	Toxicity		Organic Matter
Manganese	Treatability		Ammonia
	BOD		Heavy Metals
	TOC		Conductivity
	UV Absorption		Oil
			Gross Pollution

(c) <u>Instrument and System Requirements</u>

It is now evident that complete instrumentation systems will be required, capable of interfacing to existing and proposed telemetry systems. In general, these systems should comprise three major components or building blocks. These are:

(i)   A sample conditioning system, in most applications.

(ii) A robust, reliable and comparatively inexpensive sensor system.

(iii) A microprocessor based electronic system capable of operating and controlling the sensor and sample conditioning units, providing additional control outputs and providing alarms and data outputs to telemetry systems (e.g. RS232).

Such a system would be particularly useful in the protection of intakes to water treatment plants since in the UK about 30 per cent of the required drinking water comes from lowland rivers, other major sources being ground water and reservoirs. Although lowland rivers are a comparatively cheap source of water they are more prone to pollution than are other sources.

Pollutants involved in closure of lowland intakes, according to Dobbs and Briers, (7) have been reproduced in Table 2.

Table 2:   Pollutants Affecting Lowland Intakes

Pollutant	Percentage Occurrence
Fuel Oils	40
Farm Waste	10
Sewage	10
Plating Effluent	5
Phenols	5
Other Chemicals from Industry or Road Accident	30

About half of the major intakes have bankside storage with residence times, in the absence of short circuiting, of between 1 and 10 days. The remainder have none and water typically passes through the treatment works and into distribution within 4 hours. Recent work in France (8) has gone much further than elsewhere but even so, there appears to be a requirement for sensors of adequate sensitivity and reliability for determining low level concentrations of toxic metals, oils and other hydrocarbons, low levels of pesticide and herbicide residues as well as other carcinogenic species. Recent advances in sensor technology of relevance is provided in a subsequent section.

EXISTING MEASUREMENT TECHNIQUES

Existing measurement techniques and their limitations in terms of accuracy, and reproductibility of results, overall reliability and need for maintenance and repair of the equipment utilised are discussed below.

(a) Sensor Logistics

Prolonged exposure of sensor to industrial and domestic wastewaters and to the products of the various treatment processes inevitably leads to a degradation in performance and reliability, particularly in the case of those sensors which have evolved from laboratory instruments which were designed to measure the same or a related variable but under favourable environmental conditions. Such sensors are frequently unreliable when deployed in situations that are not climatically well controlled, where gross fouling and/or growth of biological films and slimes occurs and where skilled human supervision is not readily available.

Various remedies exist but in general, they add considerably to the cost and complexity of the instrumentation. A dual sensor system has been developed (9), initially to protect intakes to water treatment plants where the maximum reliability is required. With this system,, two identical sensors are deployed, one on stream and the other on standby (immersed in a calibrating solution containing a biocide).

By transposing automatically the functions of the sensors, both on a time basis and on receipt of data indicating that a sensor reading has deviated significantly from past data trends, it is possible to detect a sudden and serious change in sample quality on the one hand or a sensor failure on the other.

(b) Existing Sensors

State of the art data in respect of the performance of commercially available sensors for monitoring flow and quality have been published previously (3, 9) Nevertheless, it seems appropriate to provide in this paper brief details of the performance of some of the more frequently utilised sensors (e.g. sensors for monitoring flow, dissolved oxygen and temperature, oxidised nitrogen, ammonia suspended matter, organic matter and pH value).

(i) <u>Flow</u>

Since in many applications, such as consent for effluent discharge and treatment plant design, it is the mass flow of pollutants that is of greatest significance rather than actual concentrations, the current situation with respect to flow measurement technology is indicated below.

In open channels, conventional techniques based on level measurement are common but have limitations in unattended operation when applied to liquids containing high concentrations of suspended and/or floating matter. Remote level sensing using pressure or ultrasonic transducers gives better results for some applications but temperature variations can cause significant errors and both ultrasonic and optical devices have been used successfully for measurement of sludge blanket level. Magnetic flow meters are being used increasingly to measure flow of heavily polluted liquids and slurries in pipes. There are instances, however, where large variations occur and the accuracy of measurement at the lower end of the scale suffers because of zero drift. It has been suggested that magnetic flow meters with smaller internal diameters than the adjoining pipework could be used for the following reasons: (a) less susceptibility to zero drift, (b) during periods where the main pipework may not be completely full, adequate measurements could still be made in the reduced sections, provided the meter was correctly located and (c) since the throughput would necessarily be higher, self-cleaning of the electrodes mounted ion the inner wall would be more effective. However, these advantages should be carefully weighed against the increase in line pressure and the possibility of blockage under high flow conditions. A technique based on measurement of turbulence induced electrical noise at online ultrasonic sensors has been shown to give data proportional to the solids concentration. Mass and volumetric flow rates can therefore be determined simultaneously on sludges at various stages in process. (10).

(ii)　　<u>Dissolved Oxygen</u> (0-100% ±1% or 0-200% ±2% of the air saturation value)

Dissolved oxygen is perhaps the most important overall indicator of pollution and is measured utilising galvanic or polarographic probes which will work reliably in flowing samples. However, both rely on the diffusion of gaseous oxygen through a hydrophobic membrane (usually polythene or Teflon), and in polluted water it is necessary to inhibit the formation of algal growths and bacterial slimes on the membrane and to calibrate the probes regularly. To avoid interferences from gaseous hydrogen sulphide it is possible to deploy a silver/silver sulphide saltgel as an electrolyte and thus avoid the possibility of sensitivity loss by poisoning. A typical D.O. probe is shown in Fig. 1.

A recent innovation developed by Leeds and Northrup Limited which is claimed to be maintenance free is based on a coulometric principle. The probe contains three electrodes: an anode or oxygen generator, a cathode at which oxygen reduction occurs and a reference electrode. The electrolyte is contained by a polythene or Teflon membrane but this is protected by a second robust silicone rubber membrane which is far more permeable to oxygen. However, it still seems vital to provide automatic cleaning daily and some form of automatic calibration.

(iii)   <u>Temperature</u> (-10 to +40°, ±0.5°C)

Temperature measurement is seldom a problem. Platinum resistance thermometers give the best results, but in many cases the long term stability of thermistors is adequate.

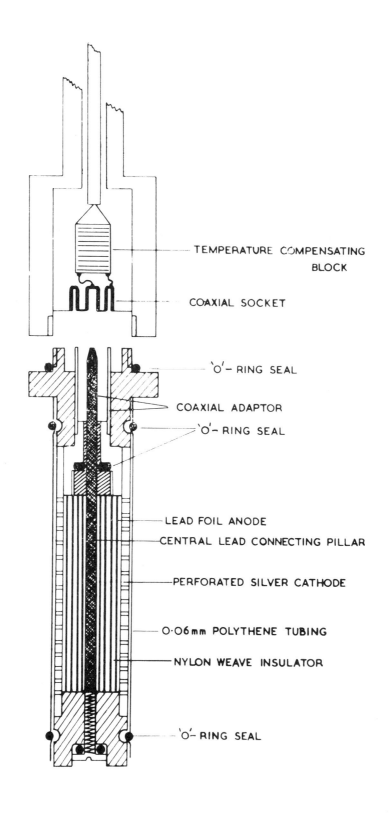

Figure 1   Schematic Diagram of a Galvanic Cell Based
DISSOLVED OXYGEN SENSOR

(iv) **Oxidised nitrogen** (0 to 50 mg N.ℓ^{-1}, ± 5% of reading)

Apart from the automatic wet-chemistry analyser approach, which is not generally recommended for unattended monitoring, the only technique known to be in common usage is the deployment of a 'specific ion monitor' fitted with a nitrate ion-selective ion-exchange electrode. The electrode performance is affected by interference from other ions in solution, in particular perchlorate, iodide, permanganate, thiocyanate and zinc. In addition, the algicide Panacide which could in other circumstances be used to inhibit algal growth, also has an adverse effect on electrode performance. Typical ion-selective monitors are shown in Fig. 2.

(v) **Ammoniacal nitrogen** (0 to 5, 10 or 50 mg N.ℓ^{-1} ± 5% of reading)

A specific ion monitor similar to that described for the measurement of nitrate ion may be used for the determination of total ammonia (as NH_3) in the sample. In this case, the electrode consists of a glass pH electrode situated behind a thin gas permeable hydrophobic membrane, with a small quantity of ammonium chloride solution in contact with the electrode tip and the membrane. When the probe is immersed in a sample containing free ammonia the latter diffuses through the membrane until the partial pressure of ammonia is equalised on both sides. Thus, as the ammonia

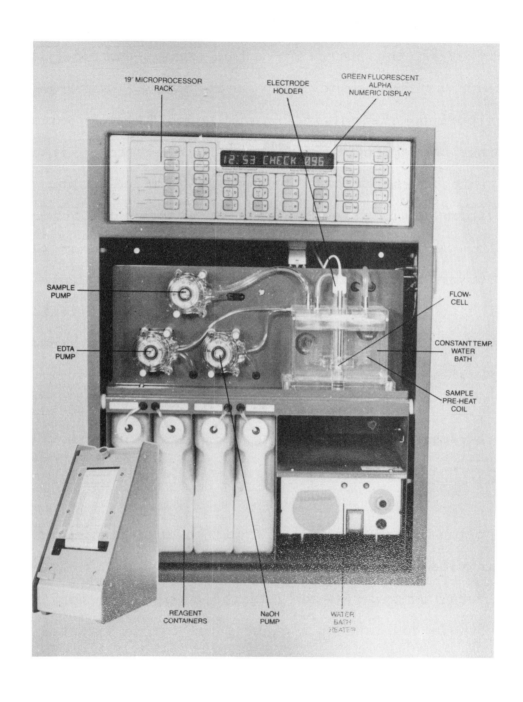

Figure 2   Photograph of a Commercially Available Ion-Selective Monitor Suitable for the Determination of Ammonia, Nitrate Ion & Fluoride

concentration in the sample changes, the pH electrode detects the change in hydrogen ion concentration and the probe as a whole has a Nernstian response to ammonia. The probe exhibits considerable resistance to contamination by dissolved ions and gases. Carbon dioxide, hydrogen sulphide and sulphur dioxide which might be expected to interfer do not affect the performance at the high pH value at which measurement of free ammonia is made. In practice, the reagent added to the sample consists of a mixture of sodium hydroxide solution to increase the pH value to 11-12 and ethylene diamine tetra acetic acid (EDTA) to minimise the deposition of hardness. A schematic diagram of the probe is shown in Fig. 3.

The response time of the probe varies from a few seconds at concentrations of about 1 mg l^{-1} to several minutes at concentration about 0.1 mg l^{-1}. When used in the specific ion monitor it will operate unattended in samples containing low concentrations in the range 0.4 to 140 mg Nl^{-1} with a precision of approximately ± 10% of reading.

(vi) <u>Organic Matter</u> (0 - 10 or 0 - 100 mg l^{-1})

Ultraviolet absorption at 254 nm has been shown to correlate well with total organic carbon (TOC) for a wide variety of samples ranging from settled sewage and effluents to raw and

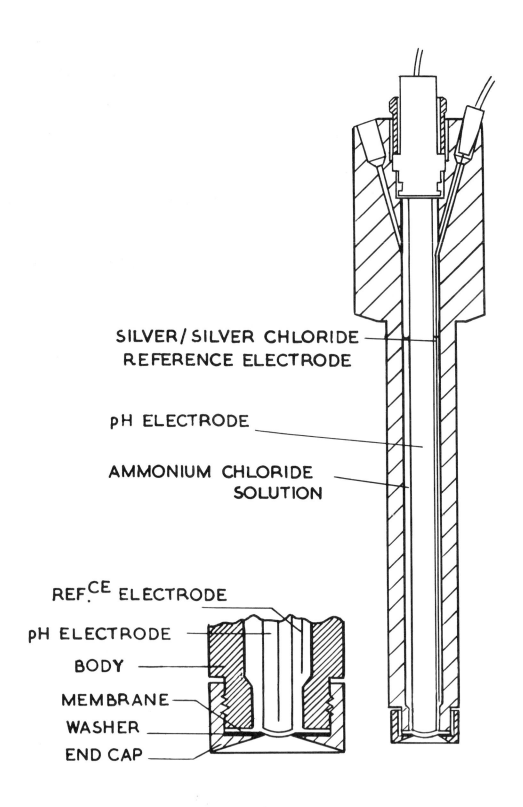

Figure 3   Schematic Diagram of an Ammonia Probe

treated river water and an instrument (Fig. 4) has been developed for water quality monitoring applications and for control of water and wastewater treatment processes. When the 'Organic Pollution Monitor' (11) first became commercially available, difficulties were encountered because of drifts in the intensities of light emitted in the ultraviolet and visible parts of the spectrum (the latter is used to correct for suspended solids which may be present in the sample). This drawback has now been overcome by the provision of a thermal stabilisation unit around the base of the UV lamp. After settling down, which usually takes several hours after initial switch on, the monitor will operate satisfactorily for at least a week without any attention.

(vii)   Suspended solids (0 to 0.1, 0 to 1.5,
0 to 500, 500 to 5000 mg l^{-1} ± 5% of reading)

Strictly speaking, this variable can only be determined gravimetrically by filtration or centrifugation of a known volume of sample followed by drying and weighing the residue. However, optical techniques have been used for some time to measure turbidity which may be related to the suspended solids concentration provided that variations in the nature of the particulates in suspension, their shape and size distribution are known (9). Turbidimeters fall into two basic categories, i.e. absorptiometers, which

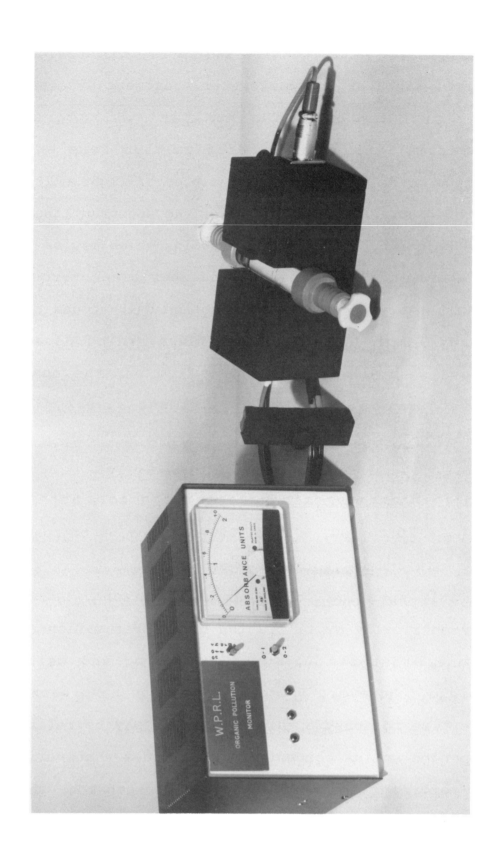

Figure 4   Photograph of Prototype Organic Pollution Monitor

measure the amount of light transmitted through sample and nephelometers, which measure the amount of light scattered at one or more angles to the incident beam. There is now a very wide range of absorptiometers and nephelometers commercially available: these satisfy most operational requirements including the monitoring of ultra-pure waters, river waters, effluents, sludges and slurries, whose concentration is measured in per cent solids (weight/volume) rather than mgl^{-1}. It should be possible therefore, by careful selection, to find a turbidimeter suitable for almost any river water quality monitoring application. Some of the better turbidimeters have built in compensation for fouling of optical surfaces and/or mechanical means of cleaning these surfaces, and in some instances the optical elements are not in contact with the sample at all. These can be calibrated by means of the Formazin standard which has been shown to be reproducible between samples and stable after several months of storage in recommended conditions.

When obtaining a practical correlation, however, it is necessary to take into account the effect of flow on particle size distribution since this will in general cause a reduction in sensitivity at higher solids concentration. Provided adequate cleaning and calibration procedures are used and the hydraulic circuit is designed to avoid trapping gas bubbles and sediment settlement, data of the required precision can be obtained.

(viii)   pH Value

A large selection of glass sensing and reference electrodes is now commercially available. To be effective, the reference electrode, which ideally should be of the double-junction type, should not be electrically grounded or present a low impedance to ground. An earthed guard electrode would be placed around the pH probe, and the glass/reference electrodes connected to a dual input stage balanced differential amplifier with a very high input impedance ($>10^{13}$ ohms). This overcomes the drift problems and provided that periodic mechanical (brush), ultrasonic or chemical (algicide and bacteriocide) cleaning techniques are also used, it is possible to monitor pH in the range 1-12 with a precision of ±0.2 pH units for periods of up to one week in an unattended operation. The interval between servicing may be extended considerably if low and high calibration solutions are fed to the probe on a daily basis and used to correct the calibration of the instrument automatically. Additionally the accuracy can also be improved.

Novel optical techniques are being developed to measure pH by monitoring colour changes in indicator dyes used to sense this variable. For example, simple devices for use in titration work have been developed (12) with low cost LED sources and work is progressing on systems where dyes are immobilised or bonded to a substrate, yet free to react with the pH sensitive medium (13). Problems of speed of response and lifetime of the probes remain, although they may well be suitable for 'one shot' operation.

(ix)   Gases

Gases which need to be monitored include oxygen, chlorine, carbon monoxide, carbon dioxide, hydrogen sulphide, methane and other flammable gases both for process control and to ensure health and safety at work.

Existing sensors are mainly electrochemical in nature and in general work satisfactorily provided they are maintained properly. However, a non-contact optically based sensor could offer considerable advantages here (3).

RECENT DEVELOPMENTS IN SENSOR TECHNOLOGY

In spite of the well published advances in microelectronics and the consequential reductions in the costs of computer and microprocessor hardware, progress with the implementation of control and automation projects in the water industry has, in many cases, been disappointingly slow. The reason often put forward for this lack of progress is the absence of sufficiently robust and reliable sensors for the measurement of those variables of significance in the monitoring and control of the particular process or plant in question. In many cases this view is only partially true and recently developed techniques such as the use of dual sensors, one on-line and one on standby, majority polling and automatic cleaning and calibration have enabled many existing sensors to be utilised successfully in a wide range of process control applications.

Frequently, however, environmental or process stream conditions are such that the requirement is for non-contact sensors or, at the very least, intrinsically safe sensors, such as those deploying fibre optics for measurement of a range of physical variables.

Unfortunately, non-contact sensing techniques are not yet practicable when monitoring such variables as treatability, toxicity and concentrations of nutrients or substrates, heavy metal ions and trace organics as would be required for example, in certain water, fermentation, pharmaceutical and medical applications. Here, the development of low and cost multiple sensor arrays based on semiconductor technology is of distinct relevance (14).

(1) **Flow Measurement**

Since a wealth of information on methods of flow measurement is readily available, particularly in respect of turbine meters, vortex meters are methods of utilising venturies flumes and weirs, this aspects will not be discussed further. However, two aspects are worthy of brief mention. These are:

a) Measurements where intrinsic safety is a major problem and/or where the environment is particularly hostile as in sewers.

b) New technological developments may well have a significant impact on flow measurements generally; these include cross-correlation techniques (10) and measurement of flow and or pressure utilising fibre or electro-optically based sensors. (3).

A significant application of lasers and fibre optics is to flow measurement through the use of the laser (Doppler) anemometer which has been commercially available for some years but at significant cost. These have unique advantages in comparison with other fluid flow instrumentation as reported by Dantec, a major European manufacturer (16). These are :

- Non-contact optical measurement
- no calibration (no drift)
- well defined direction response
- high spatial and temporal resolution
- anemometer directional measurements

However, the conditions under which laser anemometer measurements are carried out vary considerably, hence it is essential that the optical system can be adapted to the most suitable mode and configuration, for example the use of "seeding particles" which must be small enough to track the flow accurately, yet large enough to scatter sufficient light for the proper operation of the instrument.

Other fibre optic analogues of "conventional" flow meters have been proposed and discussed by Medlock (17) and may be valuable where intrinsic safety aspects are important e.g. within sewers or in the vicinity of sludge digesters where flammable gases can be a hazard and require monitoring.

(ii) <u>Sensors for Monitoring Specific Pollutants</u>

Summary descriptions are provided in subsequent paragraphs of sensing systems based on solid state sensors and electro-optic techniques (18, 19). In principle, these are cheaper to fabricate and install, frequently more useful and of more general applicability than conventional sensors.

(ii.1)   Solid-State Sensors

One of the most attractive possibilities for measuring such variables as heavy metal ions and trace organics is the deposition of appropriate sensitive materials on the gates of an array of metal oxidified effect transistors (MOSFETS) which are integrated with a microprocessor on to a single chip. Clear advantages are low cost if the market allows volume production, miniaturisation, the ability to utilise a good deal of redundancy and thus obtain better reliability and in some cases the ability to deploy in-situ cleaning and calibration. Additionally, if ratio measurements are satisfactory, then these technologies are particularly useful.

The amount of research and development going on in this area is considerable and will only be possible to discuss a few examples here. However, centres where work of this nature have been indicated and a limited number of references given. (20-24).

a)   For Monitoring Heavy Metals - An attractive approach here as pioneered at the Standard Telephone Laboratories at Harlow is the deposition on the gates of MOSFET arrays soluble phosphate and/or borate glasses containing heavy metal oxides and in consequence forming an array of solid state ion selective electrodes. If one of the gates is sensitive to a conservative ion of little interest, then the outputs of the other gates can be expressed as a ratio of the first and an external reference electrode is unnecessary. If only trends in relative concentrations are required then there is no need to make the activity of the reference ion constant in the sample and many ions could be monitored utilising a single probe incorporating the MOSFET array and the microprocessor elements needed to solve the resultant simultaneous Nernstion type equations relating ionic activities to gate potentials. Clearly, for this to be practicable then each gate must be considerably more sensitive to one ion than to the others. Early work has indicated that lead, copper, mercury, cadmium, zinc, iron and possibly aluminium and manganese could be monitored in this way but not yet with sufficient sensitivity in all cases.

An alternative approach, developed initially for measurement of pH value, sodium, potassium and chloride ions in medical applications is the use of chalcogenate glasses. Work in this area has been pioneered by Owen at Edinburgh University and Covington and Dobson at University of Newcastle upon Tyne. Basically, however, the technique is the same as that described above only the sensitive material applied to the gate being different.

Initially, however, a major problem has been that of waterproofing all areas of the MOSFET other than the sensitive material applied to the gate. Use of silicon nitride rather than silicon oxide as the base material has helped here, but there are sealing problems to be solved if the required long term stability is to be obtained. An example of a single element ion-selective field effect transistor (ISFET) is shown in Fig. 5.

b)   For Measurement of Trace Organics - Here much of the pioneering work on the immobilisation of enzymes on gates of MOSFETS has been carried out at Case Weston University and at Utar, and allied work is occurring in the UK at a number of centres, including those mentioned earlier and also at Thorn-EMI's Central Research Laboratories and at UMIST and Imperial College.

An interesting alternative is that described by Mosbach and Dannielson (25). Here enzymes are immobilised on thermistors which are used to measure heats of reaction and clearly other heat sensitive devices forming part of an integrated circuit could be utilised and the type of total fabrication suggested earlier could be deployed, once again providing sensor redundancy and data processing capability

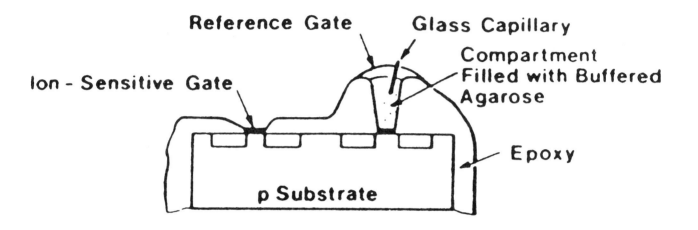

Figure 5   An Example of a Single Element Ion-Selective
               Field Effect Transistor with an Inbuilt Reference
               Electrode (from Sibbald 1986)

in a single probe. An indication of achievable sensitivity and selectivity is provided in Table 3 below.

Table 3: Achievable Selectivity and Sensitivity of Detection of a Technique Based on Measurement of Heat of Reaction Utilising Enzymes Immobilised on Thermistors or Similar Devices.

Substance	Immobilised biocatalyst	Concentration Range (mole/l)
Heavy Metal ion	Urease (eg PB^{++})	10^{-6} (detection limit)
Insecticides (e.g. Parathion)	Acetyl Cholinesterase	5×10^3 (detection limit)
Cyanide (substrate)	Rhodanase	0.02 - 1
Phenol (substrate)	Rhodanase	0.01 - 1

(ii.2)   Optical Methods of Measurement

The use of electro-optics and optical fibres coupled with microprocessors offers an attractive solution to many water industry problems, particularly those associated with measurement in or communication from hazardous areas where intrinsic safety considerations are paramount. Additionally the use of fibre optics and multiple light sensor arays coupled with pattern recognition techniques can be particularly useful as a means of monitoring plant and equipment status.

Optical sensors may be defined as those devices which use ir/visible/uv radiation to detect a change in a variable to be measured, regardless of the nature of the interaction of the light with the measurand. "Open-air" (free space propo gation) of the radiation is equally relevant, though current emphasis is on the utilisation of fibre optics in sensors for convenience of measurement. In certain cases the primary interaction between the measurand and the sensor need not be an optical interaction, but the sensor must produce an optical output after further transduction, as in the so-called "hybrid" devices or optically powered electronic sensors.

Extrinsic fibre sensors use the fibre primarily as a means of guiding light through the fibre to the measurement region, where its polarisation, wavelength, phase or intensity is modulated by the measurand and guided to a receiver where signal processing occurs. Within this class are some systems (eg time-division multiplexed, shaft-encoded and "frequency-out" sensors) where the information transfer is non-intensity dependent. In an intrinsic fibre optic sensor, the fibre itself acts as the sensor, and the propagation of light within the fibre is disturbed in a measurable way; distributed or quasi-distributed sensors are important special cases within this class. There are only very limited applications of the latter class of sensors to chemical variables, although work in the field is being undertaken.

The development of optical sensors is being stimulated by advances in optical fibre technology arising primarily through telecommunications requirements. The advantages of the use of fibre optic sensors are summarised below and slowly this new technology is becoming accepted for measurement applications.

There is a vast wealth of optical phenomena that can be utilised to measure both physical and chemical variables. Research into most of these areas is being carried out on a worldwide basis and includes research into the development of robust easy to install and maintain fibres for data transmission purposes.

The advantages and applications of Fibre Optic Techniques may be summarised as: inherent electrical safety and freedom from interference from electromagnetic "noise". Additionally fibre optics are light, small and largely free from cross-talk. With the development of powerful solid state laser sources and detectors, there is considerable potential for cost reduction and high performance from small, compact systems, rather than using lasers or lamp sources, which are fragile.

It is beyond the scope of a paper of this nature to discuss the very wide range of optical phenomena used in optical sensors and the many experimental schemes suggested to utilise these effects. Several sensor schemes are shown in Fig. 6. A number of reviews exist in the literature, often focusing upon recent trends (26) specific measurands (27) or techniques (28). In 1990 a text book will be available (29) to which a large number of experts in the field of Fibre Optic Chemical Sensors have contributed. This will represent the most comprehensive review of this very wide field and below is an illustration of the breadth of the subject. For example, the following extrinsic sensor schemes may be applied: absorptiometry, reflectometry, fluorimetry, phosphorimetry, infrared and Raman spectroscopy, evanescent wave techniques, chemi and bio-luminescence. In addition, there are the use of ground state and excited state optical sensing processes.

Furthermore, there is much research in the chemistry of sensor materials needed to make them suitable for incorporation in fibre optic devices. The use of immobilisation techniques and fluorescent labelling have been discussed. pH sensors are widely needed and a number of fluorescence and absorption based schemes proposed (13). Again, considerable work will be required before systems meeting the acceptable quality criteria of response time, use of solid state optoelectronics and avoidance of loss of

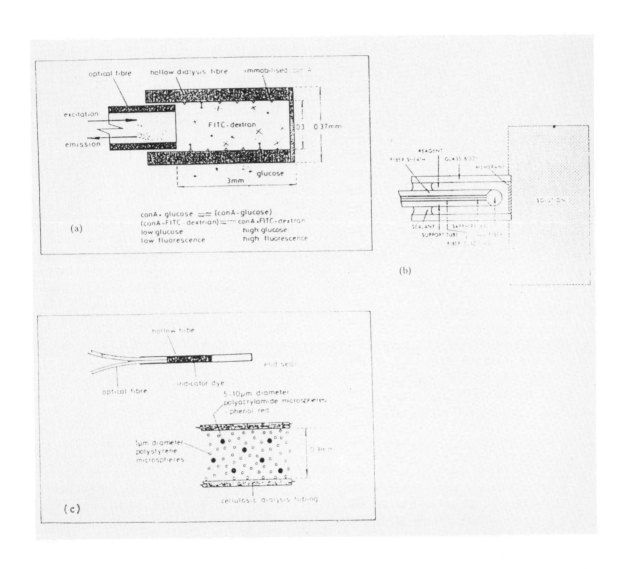

Figure 6   Examples of Fibre Optic Based Chemical Sensors (from Briggs and Grattan 1989)

the sensor material are available. Processes using changes in the characteristics of materials, e.g. cation, detection using ion-selective binding with optical readout or conversion of non-fluorescent ligands to fluorescent complexes or the synthesis of compounds which undergo colour changes upon ion-binding can be used. Dissolved oxygen is known to quench fluorescence and this may be utilised as the basis of a sensor. Chemo-retention techniques may be used where selective binding events associated with receptors located within liquid matrices can change the electrostatic and physical structure of such liquid membranes. These perturbations can be detected optically by the use of fluorescently labelled receptors and also by incorporation of fluorescent probes in liquid membranes. Hence devices using wavelength, intensity or lifetime data can be devised.

(ii.3)   Biosensors

In recent years, the possibility of deploying various biosensing techniques has attracted considerable interest. The beginnings of the use of biosensors for monitoring the process aspects of environmental protection and also source and receiving waters, were largely based on the observation of species diversification. This activity was manually intensive and included such things as the observation of the presence and proliferation of lichens whose abundance is related to atmospheric pollutants such as sulphur dioxide, sulphur trioxide and oxides of nitrogren. In the aquatic environment the diversification of species, normally resident in rivers and other water courses are almost inversely related to the degree of pollution, for example rivers with a high average dissolved oxygen level and very low levels of heavy metals and pesticide and herbicides residues or contamination by residues of organic or pharmaceutical manufacturing processes will support surface swimming fish, a wide range of invertebrates and a diverse plant population. These inhabitants slowly decrease in numbers and diversification as pollution increases culminating with anaerobic conditions where only certain bacteria and some invertebrates are present. Probably the most notable attempt to harness these phenomena to management requirements, certainly on a 'real time' automation basis was the use of Rainbow trout or other fish species to indicate the presence of toxicants in lowland rivers used as a source of potable water (30-32). In principle these devices depend on the measurement using non-contact electrodes of the minute potentials resulting from general muscular activity associated with heart activity, breathing and swimming, normal activity patterns being compared with abnormal and used to trigger an alarm. Data on sensitivity to various toxicants and system details can be found in the literature (32).

An alternative approach which also has relevance in the obtaining of early warning of potential malfunctioning of waste-water treatment plants is the use of a nitrifying filter fed by nutrients and an alequot of river water or effluent and the measurement of ammonia concentrations in the nitrifying filter effluent (33).

At a higher level of sophistication, invertebrates, whole cells and enzymes have been used in various ways. As stated earlier, Mosbach and Daniellson for example have immobilised enzymes on various thermistor configurations and related the measured heats of reaction to the degree of toxicity pertaining when a range of pollutants interact with the devices. An alternative approach which would be attractive for situations where intrinsic safety is paramount would be the immobilisation of the enzymes on an optical-fibre based thermometer.

Also of considerable interest is the work being carried out in England under the auspices of WRC at Luton College of Higher Education with assistance from Cranfield Institute of Technology and reported by Rawson and Willmer (34). Basically, the method comprises the use of mediator assisted whole cell biosensors for monitoring photosynthetic electron transfer at an electrochemical cell. To date, useful measurements have been made of concentrations of herbicide residues up stream of intakes to water treatment works. A schematic of a tangential flow electrochemical cell for on-line use is shown in Fig. 7. Results obtained to date using cyanobacterion Synechoccus have shown detection levels of less than 200 ppb with response times of less than ten minutes for selected herbicides.

More recently, a novel approach that has been put forward and is being investigated at City University, is the use of Rotifers - an invertebrate which is normally mobile and "freezes" when threatened. Here a cell containing the creatures would be optically scanned and presence or absence of movement measured and related to the presence and indeed concentration of toxicants.

Also under investigation is the use of shoaling fish or other species to give an enhanced response. These indicators or indeed others relating to a particular process could be utilised in a "dual sensor" mode. Two flow cells would be utilised one on stream and one on standby.

On receipt of an abnormal signal from the one on stream the flow cells would be transposed and the operation continued. The result would be an alternating current output where frequency and amplitude would be related to the duration and intensity of the polluting load.

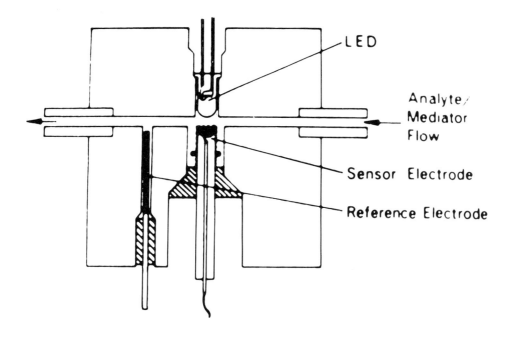

Figure 7 Tangential Flow Electrochemical Cell for On-Line Determination of Herbicide Residues (from Rawson & Willmer 1987)

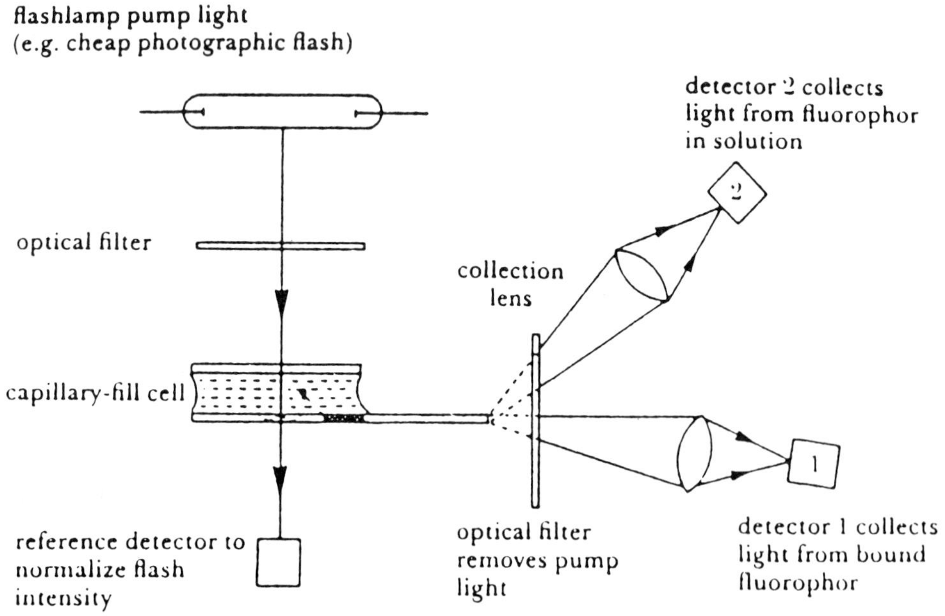

Figure 8 A Typical Instrument for Use with Fluorescence Capillary Devices (FCFD) (from Bradley et al 1987)

An optical biosensor (35) developed for immunoassay work appears to have considerable potential, initially as a "one shot" field testing device and perhaps, with further development, could form the basis of continuous monitoring devices. The device consists of two pieces of glass, separated by a narrow gap approximately 100 μ. The lower plate is coated with a layer of specific antibody and acts as an optical wave guide. The other plate has a dissoluble reagent layer of antigen labelled with a fluorescent dye. When the sample is presented to one end of the "fluorescence capillary fill device" (FCFD), it is drawn into the gap by capillary action and dissolves the reagent. If the device is set up for competition assay, the fluorescently labelled antigen will compete with sample antigen for the limited number of antibody binding sites on the waveguide solid phase. When the labelled reagent device is optically excited by a suitable wavelength light source, it fluoresces at a longer wavelength. Reagent molecules that remain in solution can only fluoresce into relatively large angles, measured relative to the plane of the lower wave guide, and hence emerge at large angles to the end of the guide according to Snell's Law of Refraction, whereas reagent molecules that are bound to the surface of the plate will emit light into all angles within the waveguide. Hence, by measuring the intensity of the fluorescent light emerging at smaller angles to the axis of the guide, it is possible to assess the quantity of reagent bound to the surface.

Measurement of the signal at larger angles can give a normalisation of the total fluoresce and hence help to compensate for background, instrumental and other sources of error. A schematic of typical instrument that could be used in conjunction with FCFDs is shown in Fig. 8.

GAS PHASE MEASUREMENTS

Measurements made in the gaseous phase can frequently be useful in water pollution control as well as in air pollution monitoring applications. Trace organics for example which have a significant vapour pressure can be measured either in the head space above large bodies of water, water courses and within water distribution and sewerage systems. Additionally, by using a "sample scrubbing" system within on-line monitors a wide range of variables can be monitored using both solid state and fiber optic electrotechniques. An example of each is described in summary form below.

Recent work by Cranny and Atkinson (36) with metal based phthalocyanine organic semiconductors deposited both as thin (thermally evaporated) and thick (screen printed) films has shown that both forms of the material have potential as gas sensors. By utilising an array sensor consisting of several individual phthalocyanine sensors on a common substrate unique responses to specific gases can be obtained. For example copper phthalocyanine can be used to monitor ammonia as NH_3 and lead phthalocyanin is responsive to various oxides of nitrogen, particularly NO_2. However, the arrays must be heated to an optimum temperature different for the thick and the thin film devices in order to maximise sensitivity and speed of response. Devices of this type could in principle prove to be useful alternatives to both the electrochemical and "wet" chemistry based in instruments currently in use.

Additionally, both active (37) and passive (38) optical devices have used from aircraft, satellites and at ground levels for measuring both air and water pollution, and portable mass spectrometers have become available for mounting on aircraft and used for monitoring environmentally polluting gases. Interest is being shown at present in the possibility of using such devices in permanent monitoring stations up stream of intakes to water treatment plants.

Also of interest in air pollution and possibly water pollution studies is the passive remote gas detector, developed initially at the Admiralty Research Establishment (ARE) and known as OTIM which is the acronym of Optical Transform Image Modulation. The technique which emulates the demodulation and filtering processes of radio receivers and performs signal processing of "gas presence" information in the optical domain, producing a Michelson interferometer type image which can be viewed by a simple photo-detector. Further developments are being carried out under an Industrial Sponsor Development Programme *. In water pollution applications, particularly at water and waste water treatment works the technique could be modified to operate with an active light source also; typical applications are being monitoring of gaseous chlorine and methane, and possibly ammonia in the headspace above inlets to sewage treatment plants.

---

* Details obtainable from Defence Technology Enterprise Ltd, ARE, Portsdown, Portsmouth, PO6 4AA.

CURRENT TRENDS IN ICA SYSTEMS

Finally, in conclusion, trends in the UK and indeed more generally, in ICA systems installed or about to be installed are summarised below:

Increased use of supervisory control and data acquisition (SCADA) systems by the recently formed Water Authority plcs and the National Rivers Authority (NRA), by industry for compliance with effluent standards and by similar bodies elsewhere. In general these systems now comprise a network of remote stations linked to one or more control stations, all for the most part based on personal computers (PCs) of adequate capability and standard programmable logic controllers (PLCs). A recent example is that described by Watts, Evans and Molloy (39) and an earlier system which deploys two minicomputers, one in "Hot Standby" mode as the control station has been described by Ramsden & Briggs (40). However, there is an almost universal trend in the direction of the deployment of SCADA systems based on a network of personal computers (PCs) since these have a number of advantages:

1) Simplified programming - much can be down-loaded to remote stations and, in many cases, common programs can be utilised.

2) Easy expandability - necessary to meet the ever changing needs of the water and water using industries.

3) Increased reliability - multiple back-up and redundancy can be deployed, hence the elimination of catastrophic failures such as can result with centralised control.

Deployment of "Easy Care" instruments and "fail safe" systems which provide a remote indication when their automatic cleaning and calibration facilities require attention. Such systems have been described previously by Briggs et al (14) and more recently, by Watts, Evans and Molloy (39).

The increased use of dynamic models in the feed-forward control of treatment processes, an example being control of coagulation and precipitation in water treatment plants. Here the commonly utilised flow proportional control of reagent addition can be improved by utilising an algorithm based on the turbidity, colour and organic content of the source water to modify the flow proportional control and a feedback element based on water quality after the settlement stage.

Improved management strategies based in many cases on "Expert System" technology. An example of such an approach is the control of biological phosphorous removal, based on both measured data and on a logical interpretation of operational practice (41).

Development of sensors and support systems for monitoring variables that can not be measured reliably on-line or close-to-line at the present time or indeed measured at all. In the above context the use of inferential measurements has been found to be of value, for example, the utilisation of redox potential measurements to control denitrification and effluent turbidity measurements to control phosphate precipitation (42).

REFERENCES

1. DOE/NWC Standing Technical Committee Report No. 27, (1981), Final Report of the Working Party on Control Systems for the Water Industry, ISBN0 90 1090 093, N W C 1, Queen Anne's Gate, London, SW1H 9BT.

2. Elvidge, A F (1989) Process Control Instrumentation in Institution of Water and Environmental Management Handbook, 1990 Issue.

3. Briggs, R and Grattan K T V (1990) Report on Measurement and Sensing of Chemical Species in the Water and Water using Industries. Obtainable from the laboratory of the Government chemist, Queens Road, Middx, TW11 0LY, UK.

4. Drake, RAR (ed) (1985), instrumentation Control and Automation in Water and Wastewater Treatment and Transport Systems, Advances in Water Pollution Control, Pergammon Press, Oxford.

5. I A W P R, (1978) 'Instrumentation and Control for Water and Wastewater Treatment and Transport Systems'. Progress in Water Technology 9 (5/6).

6. I A W P R C, (1981), Practical Experiences in Control and Automation of Wastewater and Treatment and Water Resources Management, Progress in Water Technology 13 Nos 8-12.

7. Dobbs, A J and Briers, M G, (1988), Water Quality Monitoring using Chemical and Biological Sensors, August, 1988, Analytical Proceedings 25, p278.

8. Philpot, J M, Coultant, J P and Mousty P, 1988, "Design of Accidental Pollution Alarm Systems" in Water Science and Technology, 21 Nos 10/11 1989 ISN023-1223.

9. Briggs, R (1981) Water Quality Monitoring and Control, in "Developments in Environmental Control and Public Health" ed A. Porteus, 2 p155, Applied Science Publishers, Barking.

10. Balachandran, W and Briggs, R (1981), Ultrasonic Sensors for Monitoring Flow and Sludge Solids Concentrations in Water and Wastewater Treatment, paper presented to IAWPR Workshop on Requirements, "Applications and Practical Experiences of Control and Automation in Water Quality Monitoring", Munich and Rome, June 1981.

11. Briggs, R Schofield, J W and Gorton, P A (1976) Instrumental Methods of Monitoring Organic Pollution, Wat. Pollut. Control, 25 47-57.

12. Benaim N, Grattan K T V, Palmer A W (1986) Single Fibre Optic pH Sensor for Use in Liquid Titrations. Analyst 111 1095-7.

13. Narayanaswamy, R and Sevilla III, F (1988) Optical Fibres for Chemical Species, J Phys. E. Sci. Instrum. 21, 10-17.

14. Briggs, R Meredith, W D and Solman A J, (1985) Developments in Sensor Technology in "New Technology in Water Services", 20-21 February 1985. The institution of Civil Engineers, London.

15. Medlock, R S (1982) The Techniques of Flow Measurement, Measurement and Control 15 (12), 458-462.

16. Dantec Instrumentation (1983) "Laser Doppler Anemometry" Manufacturers' Handbook and Publicity Dta, Pub: Danktec Elcktronik, Skovlunde, Denmark.

17. Medlock, R S (1986) A review of modulation schemes or optical fibre sensors. Int. J Opt Sens. 1 43-68.

18. Webb, B C (1985) Chemically Sensitive Field Effect Transistors, Paper presented at Analysis 85, Conference on Chemical Sensors, 29-30 October, 1985, from Thorn EMI Microsensors, Bury St, Ruislip Middx, HA4 7TA, UK.

19. Williams, D E, 4 Norris, J W (1987) Optical Techniques for Chemical Measurement, in Control for Success, p97 Inst. M C 87 Gower St, London, WC1.

20. Bergveld, P (1970), Development of Ion-Sensitive Solid-State Device for Neurophysiological Measurement IEEE Trnas, BME/17, 70.

21. Caras, S and Janata, J (1980) Field Effect Transistor Sensitive to Penicillin, Anal Chem 52 1935.

22. Sibbald A (1986), Recent Advances in Field Effect Chemical Microsensors J Molecular Electron, 2 51-83.

23. Sibbald, A (1985) A Chemical Sensitive Integrated Circuit: The Operational Transducer, Sensors and Actuators, 7 No 1, 23-28.

24. Sibbald A, Whalley P D and Covington A K (1984) A Miniature Flow-through Cell with a Four function Chem FET Integrated Circuit for Simultaneous Measurement of Potassium, Hydrogren, Calcium and Sodium Ins. Anal Chim Acta 159, 47.

25. Mosbach, K and Danielsson B (1981), Thermal bioanalysis in Flow Streams: Enzyme Thermistor Devices. Anal Chem. $\underline{5}$ No 1, 83A.

26. Culshaw, B (1986) Trends in Fibre Optic Sensors Int. J Opt Sens $\underline{1}$ 327-37.

27. Grattan, K T V, (1987) Recent Advances in Fibre Optic Sensors Measurement J Int. Meas. Confed $\underline{5}$ 122-34.

28. Pitt, G D, Estance, P Neat, R C, Batchelor, B A, Jones, R H, (1985) Fibre Optic Sensors, IEE Proc $\underline{1321}$, 214-47.

29. Wolfbeis, O S (Ed), (1989) Handbook of Fibre Optic Chemical Sensors CRC Press Boca Raton, Florida USA, (to be published).

30. Miller, W F (1977) Development of the WRC Water Quality Monitor Using Fish. Proceedings of the WRC Seminar on Practical Aspects of Water Quality Monitoring Systems, Stevenage, 7 December.

31. Morgan, W S G (1977) Biomonitoring with Fish: An Aid to Industrial Effluent and Surface Water Quality Control, Progress in Water Technology, 2, (3), 703-711.

32. Sloof, W (1979). Detection Limits of a Biological Monitoring System Based on Fish Respiration. Bulletin of Environmental Contamination and Toxicology, $\underline{23}$, 517.

33. Holland, G J (1977) Water Quality Monitoring in the Severn-Trent Area. Proceedings of the WRC Seminar on Practical Aspects of Water Quality Monitoring Systems, Stevenage, 7 December.

34. Rawson, D M and Willmer, A J (1987) The Development of Whole Cell Biosensors for On-Line Screening of Herbicide Pollution of Natural Waters in "Toxicity Assessment". An International Quarterly $\underline{2}$, 325-340, John Wiley & Sons Inc.

35. Bradley, R A Drake, R A L, Shanks, I A, Smith A M and Stevenson D R, Optical Biosensors for Immunoassays: The Fluorescence Capillary-fill Device, (1987) Phil. Trans. R Soc. London $\underline{B316}$, 143-160.

36. Cranny, A W and Atkinson J K, (1990) An Investigation into the Viability of Screen Printed Organic Semiconductor Compounds as "gas Sensors". "In the Press" Department of Electronics and Computer Science, University of Southampton.

37. Paris European Space Agency SP216 (1985). Remote Sensing Applications in Civil Engineering, Proceedings of a Postgraduate Summer School held at the University of Dundee 19 Aug - 8 Sept, 1984.

38. Dittman R D and Hallett, R D (1988) Remote Sensing of Phytoplankton by Laser Fluorosensors, App. Opt. 27 (15), 34290-34294.

39. Watts, Evans and Molloy (1990), A Wastewater Process Monitoring System Allowing Operating Process Control Developed by an Operational Manager, Paper to be presented at 5th IAWPRC Workshop on Instrumentation, Control and Automation of Water and Wastewater Treatment and Transport System, Yokohama and Kyoto, Japan. July/August 1990.

40. Ramsden, I and Briggs, R (1985), A Comprehensive Divisional Telemetry and Communication System, in "Instrumentation and Control of Water and Wastewater Treatment and Transport Systems", Ed R A R Drake, Advances in Water Pollution, Pergamon Press, Oxford, New York.

41. Kritchen, D J, Wilson K D and Tracey K D (1990) Application of Expert System Technology to Control of Biological Plant, Paper to be presented at 5th IOAWPRC Workshop on Instrumentation, Control and Automation of Water and Wastewater Treatment and Transport System, Yokohama and Kyoto, Japan July/August 1990.

42. Kayser (1990), Process Control and Expert Systems for Advanced Wastewater Treatment, paper to be presented at 5th IAWPRC Workshop on Instrumentation Control and Automation of Water and Wastewater Treatment and Transport Systems. Yokohama and kyoto, Japan. July/August.

ISFET-BASED CHEMICAL SENSORS FOR ENVIRONMENTAL MONITORING

J.A. Voorthuyzen
University of Twente
P.O. Box 217, 7500 AE Enschede
The Netherlands

ABSTRACT

The ISFET, originally developed for use as a liquid-pH sensitive device, has become the basic structure for a large variety of other sensors, such as enzyme- or immuno-modified ISFET's, K^+, Ca^{2+} sensitive device etc. In this paper several of the different types of ISFET-based sensors will be discussed with respect to their operational performance and possible application in controlling environmental parameters in a liquid surrounding.

Comparing ISFET-based sensors and glass electrodes, it can be concluded that due to their different way of operation as well as fabrication: ISFET's exhibit a much faster response to changes in the chemical environment, can be made with much smaller dimensions and better rigidity, can be produced in much larger volumes for a lower price, but however are troubled with a worse long-term stability.

Initial applications of ISFET's have focused on their much smaller dimensions, monitoring parameters that can not be detected by means of the much larger glass electrode, as is the case with in vivo measurements of vital parameters like blood pH. Nowadays increasing attention is paid to enzyme- or immuno-modified ISFET's as well as to dynamic applications like flow injection analysis, in order to exploit their much faster response to changes in the chemical environment and to escape from their worse long-term stability.

Typical examples of different ISFET-based sensor types are presented and discussed.

© 1990 IOP Publishing Ltd

1) INTRODUCTION

ISFET's and ISFET-based sensors have received considerably interest from researchers as well as applicants for more than fifteen years. The ISFET, originally developed for use as a liquid-pH sensitive device, has become the basic structure for a large variety of other sensors, such as enzyme-modified ISFET's and ion-sensitive device.

Due to the fact that the various types of ISFET-based sensors are diverging with respect to their fundamental way of operation and construction it is hardly possible to review their performance in a more or less general way. In this paper, therefore, several types of ISFET-based sensors will be discussed with respect to their possible application in controlling environmental parameters. An extensive review of the state of the art of almost all types of ISFET-based sensors can be found in the literature [1].

After a short introduction of the solid-state MOS transistor, the operation of the pH-ISFET, the K^+-ISFET and the enzyme-modified ISFET will be discussed. The dynamic operation of ISFET-based sensors will be explained and their performance will be reviewed with respect to their use in environmental monitoring.

2) FUNDAMENTALS OF OPERATION

2.1) The MOSFET

ISFET-based sensors are derived from their solid-state counterpart, the Metal Oxide Silicon Field Effect Transistor (MOSFET). The MOSFET is an electronic component with four electric terminals, called gate, bulk, source and drain, and operates in such a way that a relatively large electric current from drain to source may be controlled by a small voltage between gate and bulk terminal, providing a considerable amplification of electric signals. Simply the drain-source path may be considered as a resistor, which value is influenced by the gate-bulk voltage.

The physical structure of the MOSFET is drawn in figure 1. The largest part is the bulk of the MOSFET, which consists of lightly ($10^{15}/cm^3$) boron-doped silicon. Into the bulk the source and drain are realized by inversely heavy doping ($10^{19}/cm^3$) two regions with phosphorus at a small distance from each other (10-20 μm). The whole structure is coated with an insulating layer of silicon dioxide, which on top is covered with a thin conductive layer of aluminium. Contacts to the source, drain and bulk are also made by means of a thin layer of aluminium. Applying a positive voltage to the

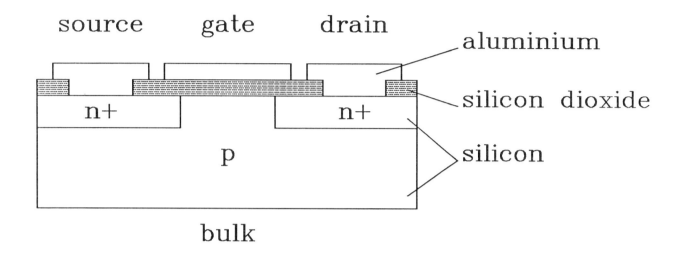

Figure 1 Cross section of the MOSFET

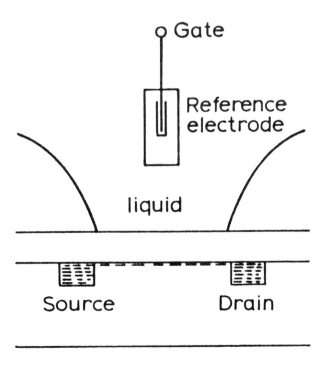

Figure 2 Schematic drawing of the ISFET

gate, electrons will be attracted to the silicon-silicon dioxide surface, causing a conductive path between drain and source. Applying a negative voltage, electrons are forced away from the surface, causing the source-drain path to be non conductive.

The actual sensitivity of the MOSFET, defined as the ratio between drain-source current change and gate-bulk voltage change, depends on MOSFET geometry. A practical value, however, appears to be 1-10 mA/V.

In practical situations the gate-bulk voltage at which a conductive path between source and drain is created, called the threshold voltage of the MOSFET, is not equal to zero, but in the region -5 V to +5 V. The actual value of this threshold voltage now appears to depend among others, on the physical properties of the gate material. Using for instance gold instead of aluminium as the gate metal, a quite different threshold voltage will be observed. It is this threshold voltage that has appeared to play a key role in the operation of ISFET's and ISFET-based chemical sensors.

2.2) The ISFET

The ISFET can be considered to be no more than a slightly modified MOSFET. In fact the gate metal is removed and the bare surface of the insulator is exposed to the electrolyte solution to be monitored, which in turn is contacted by a reference electrode, schematically drawn in figure 2.

It now appears that using insulators like SiO_2, Si_3N_4, Al_2O_3 and Ta_2O_5, at the insulator-electrolyte interface an electrical potential is generated that is a function of the electrolyte pH. Using a reference electrode, of which the potential is pH-independent, the pH of the electrolyte influences the threshold voltage of the ISFET. Measurements have shown that the sensitivity is about 25, 40, 50 and 58 mV/pH for SiO_2, Si_3N_4, Al_2O_3 and Ta_2O_5 respectively.

Extensive research has shown that the maximum signal to be expected from those insulators is 59 mV/pH, corresponding to a so-called Nernstian response [2]. Observed deviations from this maximum sensitivity are due to the fact that the selective reactivity and the density of sites, determining the insulator-electrolyte potential, depends on the insulator material, and appears to be the lowest for SiO_2 and to be the highest for Ta_2O_5.

In practical situations the ISFET chip size, including bond pads, is in the order of 1 mm * 3 mm, while the active region where the insulator-electrolyte potential is generated/detected is in the order of 0.02

mm * 0.5 mm. Besides that, the potential appears to be determined by an interfacial reaction, causing the ISFET to be a sensor with an extremely small response time to changes in the pH of the electrolyte. In case of Al_2O_3 a response time in the order of 1 msec. has been observed [3].

A serious problem of ISFET's appears to be the long-term instabilities, due to slow changes at the liquid/insulator interface or variations of the amount of electric charge stored in the insulator. It is generally believed that the long-term drift is lowest when using Al_2O_3 as the pH-sensitive insulator. Optimization of the Al_2O_3 processing conditions has shown that a long-term drift less than 0.2 mV/hour, corresponding to an apparent change of 0.1 pH/day, can be achieved [4].

The existence of a pH-sensitive device as such is not unique for the ISFET, due to the presence of for instance the well known glass electrode, developed about a century ago. Comparing the ISFET and the glass electrode, it has been observed that in the ISFET the potential determining mechanism occurs at the liquid/insulator interface, while in the glass membrane electrode it takes place in a hydrated gel layer at the solid-liquid boundary. In the glass electrode a small reservoir of liquid with reference composition is incorporated, while the ISFET is an all solid-state device. In the ISFET the solid part contacting the liquid is a real insulator, while in the glass electrode it has to be conductive.

Due to the different way of operation as well as fabrication for both types of pH-sensitive devices it appears that:
- ISFET's exhibit a 1 msec. response time to changes in the chemical environment, while the response time of glass electrodes is in the order of several seconds [3].
- ISFET's can be made with much smaller dimensions and better rigidity.
- ISFET's can be produced in much larger volumes for a lower price.
- ISFET's, however are troubled with long-term instabilities of about 0.1 pH/day. This is bad as compared to the drift of modern glass electrodes which is in the order of 0.002 pH/day.

2.3) ISFET-based K^+-ion sensor

In order to develop ISFET-based chemical sensors derived from the original pH-ISFET, early work has concentrated on the chemically modification of surface OH-groups to reduce pH sensitivity and introduce a surface sensitivity to other specific ions like K^+. It has however been found that silylation of the ISFET SiO_2 surface, which is the first step in this development, reduces the pH-sensitivity to only 20-30 % of its original value, which is far too less [5]. Therefore, later work has focused on the direct covalent attachment of thin polymeric membranes to the original ISFET surface. In this way it has been found that such devices suffer from a considerable CO_2 interference, due to the fact that not all original surface silanol groups, causing the pH-sensitivity, can be modified by the covalent bond required for membrane attachment. Extensive research has shown that the incorporation of an hydrogel between ISFET SiO_2 surface and ion sensitive membrane yields a structure with satisfying sensitivity to the specific ion and low sensitivity to interfering species like CO_2 [6], [7].

In this way a K^+- sensor with a response of 53-56 mV/pK^+ has been realized. The approximated detection limit appeared to be about 10^{-5} mol/liter and the sensitivity to other interfering ions like Na^+ has been measured to be about 10^3 times lower. A serious problem however appears to be the long term stability of such sensors as well as the design of solid-state reference electrodes.

2.4) ISFET-based enzyme sensor

A quite different approach to use ISFET's, is to cover the original pH-sensitive ISFET with a membrane in which an enzyme, like urease, is immobilized. Addition of a substrate (urea) containing liquid causes a local pH change in the membrane that will be measured by the incorporated ISFET. It has been found that a sensor with a lower detection limit of about 0.2 mmol/liter and a maximum detection limit of about 10 mmol/liter can be achieved, however with a rather bad linearity and high sensitivity to changes in buffer capacity of the surrounding liquid. To escape from these problems a so-called pH-static enzyme sensor has been realized in which the local pH-change in the enzyme-loaded membrane is compensated by coulometrically generated H^+ ions. The generating current is controlled in such a way that the pH in the membrane remains constant [8]. Experimental results have shown that such sensor/actuator systems are insensitive to changes in buffer capacity of the surrounding liquid and have a linear response to the concentration of substrate to be monitored in a quite larger range. In a similar way ISFET-based glucose sensors have been developed.

3) DYNAMIC USE OF ISFET-BASED SENSORS

Initial applications of ISFET's have focused on their much smaller dimensions, monitoring parameters that can not be detected by means of the much larger glass membrane electrode, as is the case with in vivo measurements of vital parameters like blood pH and the pH in the gastrooesophageal tract [9], [10]. In these applications ISFET-based sensors are applied in a static way, and thus are troubled with a worse long-term stability as compared to the much larger glass electrodes.

Nowadays, increasing attention is paid to the use of ISFET's and ISFET-based sensors in dynamic applications. In these types of applications not the possible miniaturization of ISFET-based sensors is the dominating feature, but their much faster response to changes in the chemical environment as mentioned in section 3 [11]. Using ISFET's in dynamic applications it is inherently possible to escape from their bad long-term stability by introducing recalibration procedures.

Examples, described in literature are an:
- ISFET-based flow injection analyzer [12]
- ISFET-based titrator [13]
- ISFET-based null detector [14]

In the ISFET-based flow injection analyzer the ISFET-based sensor, for instance a K^+-ion sensor is exposed to a continuous flow of liquid with well known K^+ concentration into which intermittently a microvolume of the solution with unknown K^+ concentration is injected. In this way transient signals are generated of which the height corresponds to the difference in K^+ concentration between both solutions. Causing the thus realized changes in concentration to be fast as compared to the sensor instabilities, a reliable output signal is obtained.

In the ISFET-based titrator, around the sensitive part of the pH-ISFET, a noble metal electrode is deposited. Injecting electric current into the solution via this electrode, in a microvolume around the electrode interface H^+ or OH^- ions are generated, which cause a local change of the pH. This pH change can be monitored by the pH-ISFET. The relation between current and pH change depends on the buffer capacity of the solution, which in turn depends on the concentration of acid or base [13]. If the titrations are performed in a relatively short time, in this approach the influence of ISFET instabilities is canceled. Note that the ISFET-based titrator is no longer used as a pH sensor, but as an acid or base concentration detector.

In the ISFET-based null detector the sensor, for instance a pH-ISFET, is alternately exposed to a reference solution and the solution to be monitored. If the pH of both solutions is different a transient signal will be generated. The pH of the reference solution is changed by adding acid or base until this transient signal disappears, in which case the pH of both solutions has to be identical.

All examples of dynamically applied ISFET-based sensors have in common that the influence of instabilities in the ISFET are reduced considerably. A drawback however is that in such applications often a more complicated set-up with valves, reference solutions, electronic circuitry etc. is required, and only semi-continuous measurements can be performed.

4) APPLICATIONS OF ISFET-BASED SENSORS

Considering the most important categories of environmental monitoring: biomolecules, heavy metals, gases and salt/ion concentrations, the following should be mentioned.

4.1) Biomolecules

The measurement of specific types of biomolecules by means of ISFET-based sensors, like urea, is already described in literature [8], [12]. Characteristic for this type of detection is the presence of a membrane on top of the pH-ISFET with an immobilized enzyme causing a local pH change in the membrane that can be detected by the ISFET. In such devices the response time is no longer determined by the transduction of pH to electrical output signal, but by the diffusion of biomolecules into the membrane and the chemical reaction caused by it. A response time of no more than several seconds has been observed.

Recently an immuno-modified ISFET has been presented suitable for the detection of biomolecules which do not cause a local pH change in the membrane, but causes a change in the membrane charge, which in turn results in a different ion mobility. The ion mobility changes can be measured by exposing the sensor to ion steps. In this way human serum albumin and progresterone have been directly detected [15].

4.2) Heavy metals

Several attempts have been made to realize an ISFET-based sensor for the detection of heavy metals. Recently the preliminary results with a cadmium sensor, based on an ISFET coated with a chalcogenide glass membrane have been reported [16]. The sensitivity appeared to be about 25 mV/decade in the concentration region of 10 μmol/liter to 10 mmol/liter.

Changing the composition of the chalcogenide glass membrane a sensor for the detection of copper ions has been realized [17]. The performance appeared to be comparable with that of conventional ion selective glass electrodes applying the same chalcogenide glass.

4.3) Gases

The detection of gases by means of ISFET's is not impossible : the ISFET is covered with a hydrated layer in which the pH can change as a function of the gas concentration to be measured. The interference with CO_2 for the K^+-ion sensor, as described in section 2.3, can for instance be exploited to develop such a sensor. The required presence of a reference electrode, liquid surroundings etc. causes this approach to be too complex. Almost no examples of ISFET-based gas sensors have been presented in the literature. A more straight forward method to detect gases is the use of so-called GasFET's: MOSFET's provided with a catalytic gate metal, of which the threshold voltage responds to certain specific gases. It is beyond the scope of this paper to discuss this type of sensors. Detailed descriptions are presented elsewhere [18].

4.4) Ions, Acids, Bases

Most of the ISFET-based sensors described in literature, are developed for the detection of ions. Once a sensor has been developed for the detection of a specific ion, other ion sensors can be realized by changing the chemical composition of the membrane in accordance with the ion to be detected. In this way ion sensor arrays have been developed for the detection of potassium, sodium, calcium and chloride [19]. Recently ISFET's have been applied for the monitoring of rain, by using a set-up in which the ISFET is continuously rinsed with a solution with an original pH equal to 5. During rainfall droplets of rain water are mixed with the slowly streaming rinsing solution, causing transient pH changes that are fast as compared to the long-term instabilities of the ISFET [20].

The ISFET-based titrator, as described in section 3, is a good example of a sensor that can be used for the detection of acid/base concentrations [13] and has on a laboratory scale already been used as an analyzer in food processing applications.

5) DISCUSSION AND CONCLUSIONS

One of the advantages of ISFET-based sensors appears to be the possible miniaturization towards millimeter scale. Initial applications of ISFET's have therefore focused on the monitoring of parameters that can not be detected by means of the much larger glass electrode, as is the case with in vivo measurements of vital parameters like blood pH. In controlling environmental parameters it is however not to be expected that the size of the sensors will trouble their possible use.

Other advantages of ISFET-based sensors, related to their way of operation as well as fabrication, are the rapid response to changes in the chemical environment and the possibly much lower price. It might be expected that especially the latter will be a key point in the development of sensors for environmental monitoring.

As already mentioned, at present most statically applied ISFET-based sensors are troubled with a bad long-term stability. Therefore it is not realistic to expect that in the near future such sensors can be applied in stand-alone monitoring of environmental parameters for a period of more than 24 hours.

The performance of dynamically applied sensors is more promising, and therefore they are expected to be more frequently applied in the near future, despite the fact that such applications however require a more complicated set-up. An other promising type of applications is the use of ISFET-based sensors as a dipstick, which if necessary, can be calibrated manually.

At present most of the sensors discussed in this paper are only used on a laboratory scale for research or diagnostic applications. Recently there is a growing interest to use ISFET-based sensors in different types of applications, like food industry, biomedicine, agriculture and pollution control.

REFERENCES

[1] A. Sibbald, Recent advances in field-effect chemical microsensors, J. Molecular Electronics, vol. 2, pp. 51-83, 1986.

[2] L. Bousse, N.F. de Rooij and P. Bergveld, Operation of chemically sensitive field-effect transistors as a function of the insulator-electrolyte interface, IEEE Trans. Electron Dev., vol. ED-30, pp. 1263-1270, 1983.

[3] B.H. van der Schoot, P. Bergveld, M. Bos and L.J. Bousse, The ISFET in analytical chemistry, Sensors and Actuators, vol. 4, pp. 267-272, 1983.

[4] Ch. Arnoux, R. Buser, M. Decroux, H.H. van den Vlekkert and N.F. de Rooij, Analysis of the structure and drift of Al_2O_3 layers used as pH-sensitive material for pH-ISFET's, " Proceedings Transducers '87, Tokyo, 1987, pp. 751-754.

[5] A. van den Berg, P. Bergveld, D.N. Reinhoudt and E.J.R. Sudhölter, Sensitivity control of ISFETs by chemical surface modification, Sensors and Actuators, vol. 8, pp. 129-148, 1985.

[6] D.N. Reinhoudt and E.J.R. Sudhölter, The transduction of host-guest interactions into electronic signals by molecular systems, Adv. Mater., vol. 2, pp. 23-31, 1990.

[7] E.J.R. Sudhölter, P.D. van der Wal, M. Skowronska-Ptasinska, A. van den Berg, P. Bergveld and D.N. Reinhoudt, Transduction of host-guest complexation into electronic signals: favoured complexation of potassium ions by synthetic macrocyclic polyethers using membrane-modified, ion-sensitive field-effect transistors (ISFETs), Recl. Trav. Chim. Pays-Bas, vol. 109, pp. 222-225, 1990.

[8] B.H. van der Schoot, H. Voorthuyzen and P. Bergveld, The pH-static enzyme sensor: design of the pH control system, Sensors and Actuators, vol. B1, pp. 546-549, 1990.

[9] P. Bergveld, Sensors for biomedical applications, Sensors and Actuators, vol. 10, pp. 165-179, 1986.

[10] L. Thybaud, C. Depeursinge, D. Rouillerr, G. Mondin and A. Grisel, Use of ISFETs for 24 h pH monitoring in the gastrooesophageal tract, Sensors and Actuators, vol. B1, pp. 485-487, 1990.

[11] P. Bergveld, Exploiting the dynamic properties of FET-based chemical sensors, J. Phys. E: Sci. Instrum. 22, pp. 678-683, 1989.

[12] G.K. Chandler, J.R. Dodgson and M.J. Eddowes, An ISFET-based flow injection analysis system for determination of urea: experiment and theory, Sensors and Actuators, vol. B1, pp. 433-437, 1990.

[13] W. Olthuis, J. Luo, B.H. van der Schoot, J.G. Bomer and P. Bergveld, Dynamic behaviour of ISFET-based sensor-actuator systems, Sensors and Actuators, vol. B1, pp. 416-420, 1990.

[14] W. Gumbrecht, W. Schelter, B. Montag, M. Rasinski and U. Pfeiffer, Online blood electrolyte monitoring with a ChemFET Microcell system, Sensors and Actuators, vol. B1, pp. 477-480, 1990.

[15] R.B.M. Schasfoort, J. Bomer, P. Bergveld, R.P.H. Kooyman and J. Greve, Modulation of the ISFET response by an immunological reaction, Sensors and Actuators, vol. 17, pp. 531-535, 1989.

[16] J. Salardenne, J. Morcos, M. Aït Allal and J. Portier, New ISFET sensitive membranes, Sensors and Actuators, vol. B1, pp. 385-389, 1990.

[17] Y.A. Tarantov, Y. G. Vlasov, Y. A. Mesentsev and Y.L. Averyanov, Physical and chemical processes in ISFETs with chalcogenide membranes, Sensors and Actuators, vol. B1, pp. 390-394, 1990.

[18] I. Lundström, A. Spetz, F. Winquist, U. Ackelid and H. Sundgren, Catalytic metals and field effect devices - a useful combination, Sensors and Actuators, vol. B1, pp. 15-20, 1990.

[19] S.-I. Wakida, M. Yamane, K. Higashi, K. Hiro and Y. Ujihira, Urushi matrix sodium, potassium, calcium and chloride-selective field-effect transistors, Sensors and Actuators, vol. B1, pp. 412-415, 1990.

[20] P. Alean-Kirkpatrick and J. Hertz, An ion-selective field effect transistor for in situ pH monitoring of wet deposition, Intern. J. Environ. Anal. Chem., vol. 37, pp. 149-159, 1989.

IN-SITU CONTINUOUS FIBRE OPTIC SENSORS FOR WATER POLLUTANTS

Dr. Brian MacCraith

Optical Sensor Laboratory

School of Physical Sciences

Dublin City University

Glasnevin

Dublin 9

Abstract

A review is presented of the application of fibre optic sensor technology to the continuous monitoring of specific water pollutants. A comparison is made between this monitoring approach and the conventional diagnostic sampling approach. A range of sensor types is described, based on fluorescence, absorbance and other spectroscopic techniques. Examples are given of sensors for organochlorides, phenols, fuel oils and pesticides. Descriptions of portable commercial instruments developed in the U.S.A. are presented. In addition, recent developments in this area at Dublin City University are described.

1. **Introduction**

The success of environmental management and protection depends largely on the quality and extent of information provided by the various data gathering networks put in place by monitoring agencies. In the particular case of the aquatic environment the situation is complicated by the vast array of contaminants/water quality parameters which require monitoring and by their often difficult or remote location. Traditional sampling methods suffer from the time and man-effort involved and, in some cases, doubts about the integrity of samples when they eventually reach the laboratory. This is particularly true in the case of volatile contaminants. In addition, the sampling approach does not provide early- warning signals in cases of major contamination problems where speed of response is critical. Thus, for public health and ecological reasons, there is a critical need for <u>continuous</u>, <u>in-situ</u> monitoring of specific priority pollutants in the aquatic environment.

In this paper, in-situ, real-time monitoring capabilities based on fibre- optic sensing technology are described. This relatively new technology applies advances in fibre optics and lasers to chemistry and spectroscopy, thereby enabling remote monitoring of sub ppm quantities of specific molecules. The technique is versatile in that it is suitable for in-situ surface water, borehole and ground water monitoring. It may be used as an early warning system at known critical sites or to provide continuous monitoring of compliance with specific E.C. directives on drinking water quality, for example. The instrumentation is portable (or at least mobile, depending upon application), ruggedised and simple to operate.

2. **Sampling vs In-situ Monitoring**

    Conventional methods of water quality monitoring generally rely on sophisticated sampling techniques and expensive laboratory-based analytical instruments such as gas chromatographs and mass-spectrometers. Such approaches are diagnostic in nature in that each sample is treated as an unknown quantity and is fully characterised by the analytical procedure. In addition, skilled personnel are required to carry out this activity. The major disadvantages of the conventional approach are:

    (i) Expense, e.g. in the U.K. £34m per annum is spent on sampling and laboratory analysis. Of this, £10m is expended on sample collection activities alone.

    (ii) Turn-around time. The transport and analysis of samples may take days before results are provided.

    (iii) Doubts about sample integrity, e.g. labile or volatile contaminants may have changed their concentrations considerably by the time analysis is performed.

These disadvantages, together with the recent emphasis on compliance with national and E.C. directives on water quality, have directed attention to real-time in-situ monitoring techniques as provided by fibre optic sensing technology, for example. In contrast with the diagnostic approach, here specific sensors are dedicated to monitor particular contaminants, often with high sensitivity and specificity. Such in-situ monitoring overcomes problems of sample integrity and may be used to provide early warning alarms particularly if the sensors are coupled to conventional telemetry systems. These sensors are typically portable, versatile and simple to operate, providing a direct reading of the contaminant concentration. Cost-effectiveness is enhanced when many sensors are coupled to a single signal processing instrument.

3. Principles of Fibre Optic Chemical Sensing

As depicted in Fig.1 fibre optic chemical sensing falls into two categories:

(a) Remote Fibre Spectroscopy

Here the fibre is used as a light guide to perform direct spectral analysis (fluorescence, absorbance, raman) of a sample at a distance. Groundwater contaminants have been monitored in this manner at sub-parts per million levels over distances of hundreds of metres (Ref. 1).

I REMOTE FIBRE SPECTROSCOPY (RFS)

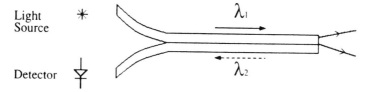

II FIBRE OPTIC CHEMICAL SENSORS (FOCS)
    " OPTRODES "

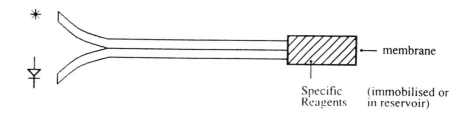

Figure 1. Principles of Fibre Optic Chemical Sensing.

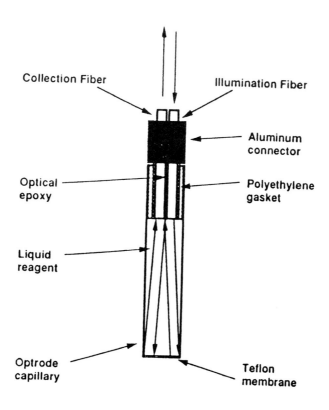

Figure 2(a). Reservoir-type absorbance-base Sensor Head.

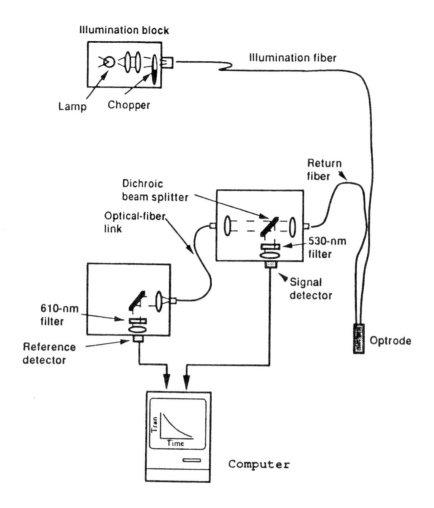

Figure 2(b). Associated optics and detection system for absorbance probe. These components are contained in a portable instrument.

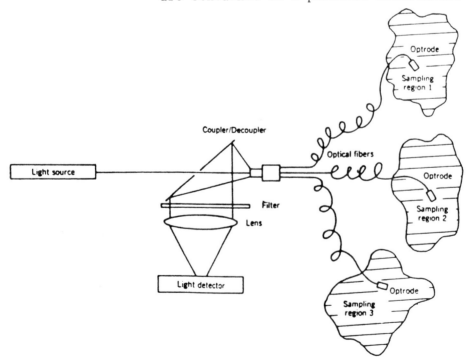

Figure 3. Schematic of remote fluorescence-based sensor.

(b) <u>Fibre Optic Optrodes</u>

Here the optical fibre is combined with specific chemistry. At the distal end of the fibre specific reagents are contained in a miniature reservoir attached to the fibre tip and are separated from the sample by means of an appropriate membrane. Alternatively, suitable reagents may be immobilized directly on the fibre tip. Reagents are chosen to react sensitively and specifically to a contaminant analyte and the resultant absorbance (or fluorescence) change is a direct measure of the analyte concentration. A reservoir type absorbance-based sensor is shown in Figure 2.

A schematic diagram of a fluorescence-based fibre optic sensor is shown in Figure 3. An instrument based on such technology can be made completely portable. At present, single-channel instruments have been developed in the U.S. (Ref. 2). Future units will be multichannel and multiplexed, thereby providing enhanced cost-effectiveness in that a single detection unit can be used to monitor a number of sites for a number of different pollutants. Fibre optic sensing technology has already achieved successful commercialisation in industrial process control (Ref. 3) and in critical care medical monitoring (Ref. 4). First generation systems applied to environmental monitoring have recently become available in the U.S. where research in this area has been pioneered.

4. **Examples of Fibre Optic Sensors for Water Quality Monitoring**

A wide range of applications of fibre optic sensing technology to water quality monitoring has been reported in scientific literature. Most research activity has been performed in the U.S. where the Environmental Protection Agency embarked on a Fibre Optic Sensing Program in 1982. The following examples illustrate many of the features of the technology:

(i) Remote Groundwater Monitoring using Laser-Induced Fluorescence and Fibre Optics (Ref. 5):

Researchers at Tufts University, Mass., U.S.A. have designed a portable laser system for in-situ monitoring of naturally fluorescing compounds in groundwater. In particular, aromatic contaminants such as benzene, phenol and toluene have been monitored at ppb levels using fibre optic cable lengths up to 30m. UV laser excitation at 266nm was used to stimulate the fluorescence in these compounds, which were distinguished by their different spectra as shown in Fig. 4. In addition, it is clear that interferences from humic acids, for example, can be distinguished using this technique. The experimental arrangement is essentially that shown in Fig. 3. Quoted sensitivities over 25m cable lengths are as follows:

> phenol   10ppb
>
> toluene   9ppb
>
> xylene   0.07ppb
>
> fuel oil 0.07ppb

This technique is particularly suited to giving early warning of fuel oil spills and to the monitoring of clean-up operations at contaminated sites.

(ii) <u>Remote Detection of Organochlorides</u>:

Chlorinated solvents such as trichloroethylene and chloroform are becoming increasingly common as water contaminants. Many of these compounds are carcinogenic and all are undesirable in drinking water. These volatile chlorinated solvents are detectable via the Fujiwara reaction (Ref. 6), whereby polychlorinated organic compounds react with alkaline pyridine to produce a fluorescent chromophore. The reaction can be monitored via fluorescence or absorbance in a reservoir-type optrode design (Ref. 7). The absorbance approach is illustrated in Fig. 2 where trichloroethylene is monitored via absorbance at 530nm referenced to a non-absorbing wavelength at 610nm. Less than 10ppb sensitivity has been demonstrated with this probe which uses a modified commercially available portable fluorimeter (Ref. 2). A particular feature of the probe is that the volatile nature of the compounds being monitored allows a headspace analysis to be used, thereby minimising the potential problem of

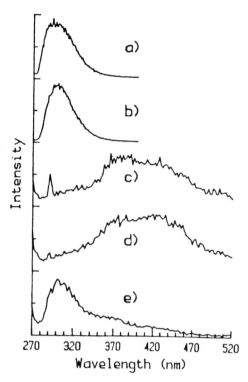

Figure 4. Fibre optic fluorescence spectra obtained for (a) phenol, (b) o-cresol, (c) humic acid, (d) bark leachate and (e) landfill leachate.

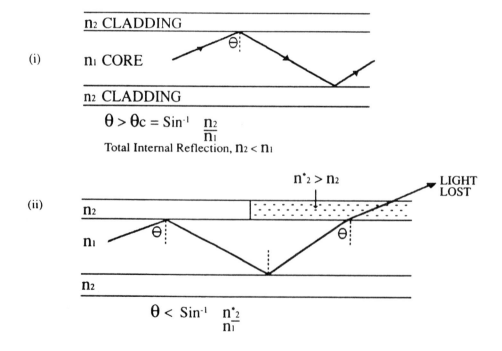

Figure 5. Principle of operation of Refractive index Fibre Optic Chemical Sensor.

fouling. Preliminary reports indicate good agreement between this optrode and independent gas chromatography/mass spectrometry measurements of collected samples. The technique may also be applied to contaminated soils.

(iii) <u>Monitoring of TCE using a Refractive Index Fibre Optic Chemical Sensor</u>:

This example illustrates a family of devices being commercialised in the U.S. by S.T. & E. Inc., California and Fiberchem, New Mexico. The principle of operation is illustrated in Fig. 5. The fibre cladding material is replaced locally by an organophilic, hydrophobic coating (typically an octadecyltrichlorosilane) which has a high specific affinity for the target contaminant. The coating absorbs the contaminant from the water and thereby raises the refractive index of the cladding. Consequently, some light which was originally guided by total internal reflection is now lost from the fibre. Thus, the transmitted signal can be related to the contaminant concentration. Using a range of coating materials different specific sensors can be fabricated. Sensors for TCE (trichloroethylene) and specific petrochemical pollutants are available. Sensitivities in the ppb region and transmission distances up to km have been demonstrated.

5. **Practical Difficulties;**

   Sufficient user information is not yet available to enable a proper evaluation of this technology. It is known, however, that certain technical aspects do require further research. These include:

   chemical immobilisation methods

   longevity of reagents in the field

   anti-fouling methods

   membrane technology

   optical fibre quality.

6. **Sensor Work at DCU:**

   The Optical Sensor Laboratory at Dublin City University has been working in the area of fibre optic sensors since 1986 and has developed techniques in the area of remote gas sensing and biosensors. In the past year work has been initiated in the field of water quality monitoring using optical fibre technology. A major collaborative project with the Analytical Chemistry Laboratory at University College Cork is being undertaken from Summer 1990.

7. **Conclusions:**

   Fibre optic sensor technology has an important role to play in the field of water quality monitoring. It offers capabilities for in-situ continuous monitoring and is particularly suited to groundwater contaminant quantitation. High sensitivities are achievable and the potential exists for enhanced cost-effectiveness over current technologies.

References:

1. Hirschfeld, T. et al., Opt. Eng. 22 (5), 527-531, 1983.

2. Douglas Instruments, Portable Fibre Optic Fluorimeter: Model II, Palo Alto, CA, USA.

3. Metricor, Fibre Optic Multisensor System, Washington, USA.

4. CDI Blood Gas Monitoring System, Cardio-vascular Devices Inc., USA.

5. Chudyk, W. et al., Anal. Chem., 57, 1237-1242, 1985.

6. Fujiwara, K., Sitzungsber, Abh. Naturforsch. Ges. Rostock., 6, 33, 1916.

7. Milanovich, F.P. et al., Anal. Instrum., 15 (2), 137, 1986.

Fibre Optic Sensors for Environmental Monitoring:
Can They Meet the Task ?

K.T.V. Grattan

Measurement and Instrumentation Centre,
Department of Electrical, Electronic and Information Engineering,
City University, Northampton Square,
London EC1V 0HB, England.

Abstract

With the increasing need for new and accurate environmental sensors, the question is raised as to whether fibre optic sensors can meet the task. In this work, several of the problems in the field are addressed and possible solutions, in light of experience with the development of physical sensors, are considered.

Introduction and Background

Many articles proclaiming the enormous benefits to various industries of fibre optic sensors have been a feature of both the technical and popular press in recent years. In, for example, one of these, early predictions of fibre optics providing "useful new measuring systems for Industry by the mid 1980's" have not been realized, as yet. The question which must be asked is as follows: is this a result of the failure of the technology to live up to the high expectations of the early reports of its development or is it a case of the reluctance of the process industries to take up the challenge of the use of new and powerful sensing systems?

The parallel with the development of the laser is one which can easily be drawn. The device that was seen as "a solution looking for a problem" could indeed have its close counterpart in the optical fibre sensor. The important feature, however, seems to be the slow development of the technological niche in which the fibre optic sensor can really show its capabilities, in much the same way that the semiconductor laser, operating at room temperature, has revolutionised the use of the laser in modern technology. Indeed the cliche "every home should have one" is almost fulfilled in the use of this device in the CD player, familiar in the modern HiFi system. It is satisfying to note that there is an increasing number of products

© 1990 IOP Publishing Ltd

available on the instrumentation market which are fibre optic based and which can offer real solutions to real industrial sensing problems. What must be considered is whether the fibre optic sensor can have a real impact upon environmental monitoring and whether the climate that exists for "green" technology is one which can be met through the use of these innovative optical sensing systems.

Optical Fibres for use in Sensor Systems

The optical fibre is a dielectric wave guide, in general of glass, consisting of a core region and a cladding region, as shown in Fig. 1. The light is guided as a result of the refractive index difference between the core and cladding and developments since the mid 1960's have ensured that very low levels of attenuation can now be achieved in optical fibre transmission. This characteristic is particularly valuable for communication purposes and often is a bonus for optical fibre sensors where considerable lengths of fibre are needed between the sensor head itself and the control and processing point. The attenuation of optical fibres is however wavelength-dependent and although for communication purposes this is rarely a significant factor, as most communication systems tend to operate at wavelengths where the optical loss is low, optical fibre sensors systems may be constrained to operate at other wavelengths as a result of the nature of the sensing process and so the attenuation becomes very significant. For example, a fibre which may show an attenuation of 0.2 dB m^{-1}, at a wavelength in the near infrared, can show one thousand times or even ten thousands times that attenuation at the edge of the ultra violet part of the spectrum. As optical measurement systems for environmental parameters frequently use such short wavelength radiation, this is an important consideration.

There are two major cases of optical fibre sensors which have been developed: _intrinsic sensors_ in which the sensing element is the optical fibre itself and more frequently _extrinsic sensors_ in which the fibres simply carry light to them from a sensor head where the optical interaction occurs. The latter class is generally inherently simpler in terms of the data processing that is necessary.

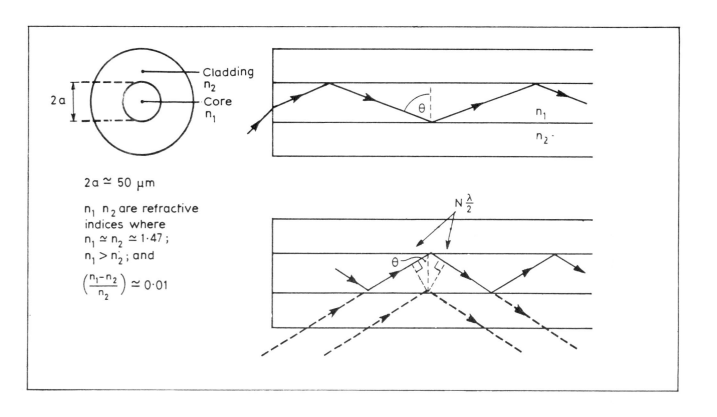

Figure 1 :Schematic optical fibre

The positive features and most widely claimed advantages of optical fibre sensor systems have frequently been listed. They include the use of an insulating material, in general silica, and the transmission itself is free from interference by surrounding radio frequency signals. Thus optical fibre sensors can be used in power stations, in areas where electrical discharges are operated, near high power radio beacons etc with no cross-talk problems. In general, sensors are of relatively low mass, and are frequently described as intrinsically safe. Questions regarding the intrinsic safety of sensors have been raised and several reports have been produced by Sira Safety Services Limited (1), and it is the opinion of the author that for most practical purposes, optical fibres using low power sources are safe even though the optical fibre may be fractured and light may "leak out" from the fibre. Further, these sensors systems are usually compatible with a recently developed optoelectronic devices such as laser sources, solid state detectors, and other components produced for and available to the optical communications industry. As a result it can frequently be said that optical fibre sensors have been constructed with a substantial part of the development cost met through the communications industry. Additional desirable features of optical fibre sensors, which are less frequently achieved are an output in a non-intensity dependent form e.g. a digital form, in terms of frequency or time, and compatibility with other sensor and communication technology i.e. readily interfaced to optical communications or digital data transmission. In spite of the fact that optical fibre sensors use optical transducers, it is often the case that a conversion to electrical signals first must be made before there can be a re-conversion to a optical data transmission, because of the analogue output of many optical fibre sensors. In addition, there is a basic requirement now that there must be a clear advantage shown over existing technology for a particular application of optical fibre sensors. To overcome the costs involved in staff who currently are trained in the use of electrical sensors to be retrained to deal with optical technology, there is a need for additional positive features in the employment of the optical techniques. This will occur, for example, where the intrinsic safety considerations or r. f. immunity are vitally important. Quite frequently this sort of situation arises where a newer and more difficult sensing need has to be met and where conventional technology cannot achieve this, at reasonable cost.

Techniques Available For Fibre Optic Sensors

A very wide range of techniques have been employed in fibre optic sensors for physical sensing and many of these could be converted for use with optical fibre chemical sensors for environmental applications. For example, for extrinsic techniques, these range from a simple shutter-type switch, operating on a basic "on/off" mode, through the monitoring of absorption and fluorescence of the material to be sensed, the use of electro-optic and polarization effects (in particular in current and voltage sensing) and the use of wavelength division encoding. This latter technique is particularly powerful as a non-intensity dependent method, in which the information can frequently be transduced by the use of a grating or interference filter which is rotated and the transmission or reflection characteristics of the device will change as a function of the angle of rotation. Therefore, if the measurand information is converted to be a function of the wavelength of light which is reflected from a grating, for example, from a white light source, then that which would strike a fixed angle detector would change. Powerful and simple encoding devices have been developed recently in several laboratories (2) in order to exploit this phenomenon, as illustrated in Fig. 2.

In addition the use of radiation which is passively emitted by a device has been frequently employed. This is most readily seen in optical pyrometry where high temperatures can be measured through the black body radiation produced by a device (3). However this could be employed in other cases where natural fluorescence or phosphorescence occurs and this can be observed and used as an indicator of the state of the material. Interferometric techniques are very powerful where small displacements may be sensed through the use of a fibre optic or an open air interferometer arrangement (4,5). Several sophisticated signal processing approaches are available in order to overcome the inherent ambiguity in the use of interferometric techniques over a wide range (6), but it is unfortunate to note that as yet very few devices (apart from the optical fibre gyroscope) employ the use of an interferometer as the primary sensor device. This arises most frequently from

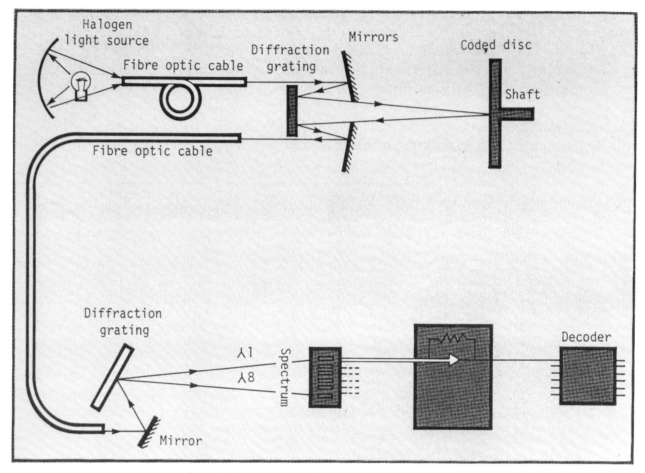

A typical example — the optical shaft encoder

Figure 2: Illustration of wavelength division modulation approach for optical sensing.

alignment. Thus their employment in the laboratory is rather different from their installation in the field.

Intrinsic optical fibre sensing devices are most frequently experienced in the measurement of temperature. The most successful of these is the optical fibre temperature monitoring system that has been recently marketed by York Technology (7). Through the use of the monitoring of the Raman scattering of light from an optical source transmitted by the fibre, and the monitoring of the time of transmission of the light in the optical fibre to a nanosecond resolution, both temperature and spatial information can be obtained. Thus an accuracy of a few degrees with a spatial resolution of a few meters can be obtained through the use of conventional silica fibre. Other variations of this technique have been reported using specialised fibres (8) (which are inherently more expensive to produce) but it does offer a unique sensing opportunity, by comparison to the use of conventional techniques. Thus one the major questions about fibre optics sensors meeting the task of environmental sensoring is whether or not such an application which is successfully met can be seen in the development of environmental sensing devices.

What Is The Nature Of The Task?

The optical fibre sensing task for environmental monitoring is many faceted, but the are several features which must be considered. It is obvious the optical fibre sensors for this (and indeed any) technique must be reliable and have low maintenance, be efficient and offer adequate accuracy and reproducibility, be available at a cost compatible with the need for the measurement itself and in general have a low energy consumption. This is particularly important in situations where larger number of sensors will be employed, where they are remote from the source itself and mains power is not available. Overall, the requirement is that optical fibre sensors are able to perform the task in a way that is better than that achieved with existing techniques, or there is the ability to make a measurement which cannot be made at all, or only with extreme difficulty, using a conventional approach. In environmental monitoring, there is growing concern that the increase of pesticides and pollutants which are released into the air can cause damage to the human health and to the natural environment on a local, national, or indeed international scale. A greater awareness of this danger has stimulated requirements for better environmental practice and the existence of international agreements and of European Community regulations has given impetus to the need for measurements in the environment which previously had not been required. These factors have in turn generated pressing needs for accurately traceable standards to calibrate existing instruments and for the development of a new, reliable, and cost effective methods for measurement. It is vital that our industry is able, in succeeding years, to demonstrate compliance with International specifications standards and with those of the European Community.

Fibre Optic Sensors For Physical Parameters
- Case Studies in Meeting The Task

Over recent years a wide variety of physical parameters have been the subject of intensive research in the development of fibre optic sensors. In particular such parameters as temperature, pressure and strain, flow and level, current and voltage, velocity, displacement, and refractive index have been the subject of a very large number of publications in the technical literature and many commercial products, some more successful than others. In addition, as discussed previously, distributed sensing techniques have been employed in order to make measurements of a particular parameter along an optical fibre with a defined spatial variation. Several comprehensive reviews have been published, e.g. (9). In considering the case of fibre optic sensors meeting a specific task of monitoring, then it is important to view how the better established field of sensing of physical parameters had been developed in recent years, and to look for technologies which, having been developed, may be employed in the monitoring of the environment.

1. Temperature Measurement

There is probably no area of fibre optic sensing which has been considered in greater detail than that of temperature measurement. A number of reviews of this subject have been written (10,11) and the many schemes have been considered. Temperature can be measured over a wide range from sub-room temperature, through the biological range to temperatures in excess of several hundred degrees and upwards. Thus situations as diverse as monitoring of temperatures of explosive mixtures at freezing temperatures, through biological probes, through to the heating of blast furnaces are being considered for inefficient combustion can lead to an increase in pollution. Of the commercial devices produced the most successful is that produced by Luxtron (12), using the monitoring of temperature-dependent fluorescence. The initial Luxtron device was the type 1000 temperature sensor, which is illustrated in Fig. 3, where the sensing element is a rare earth phosphor and light from a UV lamp source addresses this material. The longer wavelength light emitted due to the fluorescence properties of the sample is observed and the ratio of two wavelengths in this emission is determined, as a function of the temperature to be measured. In 1985, a variation of this device using the monitoring of fluorescence decay time was developed by Luxtron, sold as Model type 750. In this device, again, a UV lamp source is used to monitor the fluorescence of a material, magnesium fluorogermanate, the decay-time characteristic of which is shown in Fig. 4. The decay-time approach is a particularly powerful one because the temperature information is encoded as time information and since the work of James et al in 1979 (13), a number of devices have been reported including those where a television phosphor material was used. Work by the author with others has concentrated on the use of laser materials e.g. neodymium and ruby (14) and accurate results from a working device have been produced. In a recent report by Wickersheim of Luxtron (15) it was reported that some thirty million dollars worth of sales of these fibre optic temperature sensing devices had been achieved over the last ten years and the rate of sales was increasing. In particular, new and unique applications of temperature monitoring, where fibre optics offered a particularly valuable solution, had been seen. An illustration of this was in the monitoring of the heating of pre-processed foods in microwave ovens where a series of fibre optic probes were inserted into the food and the rate of rise of temperature was observed. Fibre optics are highly compatible with the use of microwave ovens because the interaction between the microwaves and the transducer is limited and by comparison to the use of thermocouples or other devices, a much simpler sensing arrangement is possible. Thus a particular niche market has been achieved for fibre optic temperature sensing and though the device is capable of replacing the thermocouple for routine sensing applications, it is by no means cost effective to do so. This raises the important lesson that fibre optic sensing devices are best suited to those applications where they offer particular and specific advantages and

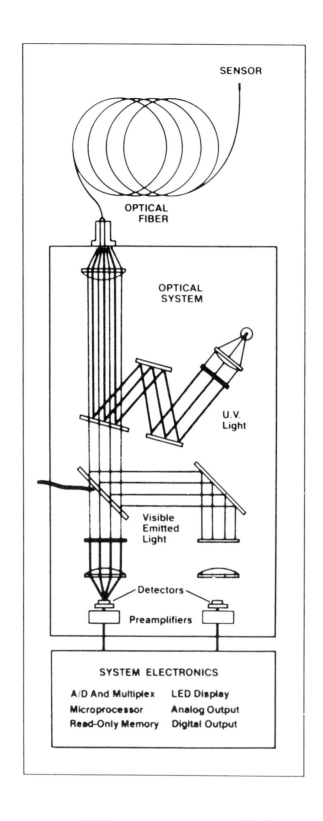

Figure 3 :Luxtron type 1000 temperature sensor device.

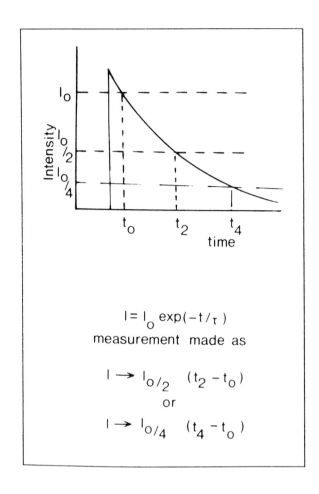

Figure 4: Illustration of 'decay-time' approach to temperature sensing.

where the inevitably higher costs can be justified on the basis of offering unique sensing applications.

Further, the monitoring of temperature in the high temperature environment is one where fibre optic devices are able again to provide a particularly valuable solution and the Accufibre temperature monitoring system, has developed by workers at NBS in the United States, is used as a temperature standard in the 630- 1060C range (3). The temperature is measured from the radiation power at two selected wavelengths. Again a unique sensing opportunity is available through the use of this device which is however a variation on the optical pyrometry which forms the basis of the international temperature standard.

2. Pressure

A number of devices for fibre optic sensing have been reported and these range in complexity from again simple shutter devices e.g. the "Fotonic" sensor which was reported some twenty years ago through those which use polarization changes, frustrated-total internal-reflection techniques, through to optical readout of Bourdon tubes to the change of resonant frequency of crystals, wires or diaphragms. This latter technology is proving particularly successful and a number of industrial companies have concentrated on the development of devices using these aspects of the technology. Vibrating element structures rely upon the modulation of a mechanical resonant element, which is essentially the transducer itself, to yield the desired information, as shown in Fig. 5. A simple change of resonant frequency can be induced and monitored by optical means and these devices are also capable of operating in the so-called 'hybrid' mode i.e. that is where the transmission of power is by optical means through an optical fibre but the information may be derived at the sensor itself by electrical means by its conversion to an optical signal and then its reconversion to a conventional signal at the control and processing box (16). A considerable amount of effort has been expended by major industrial companies in the development of these sorts of devices, both hybrid and all optical, and they offer particular solutions to sensing needs in where commercial sensors are unsuitable. However, again, there is a problem of the use of such devices at costs which are compatible with the existing technology and for example, like the thermocouple,

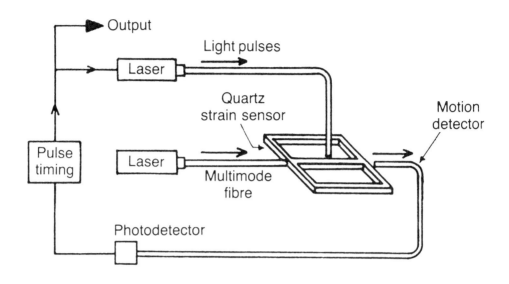

Figure 5 :Schematic of vibrating resonant element used in optical fibre sensor.

readout can address particularly difficult problems where there is considerable level of electromagnetic noise. There is considerable scope for the use of this technology in environmental sensing, by again exploiting the same phenomenon, induced by chemical and not physical changes.

3. Current/Voltage

In recent years considerable effort has been expended in the development of fibre optic current and voltage sensors and in this area, through the use of electro-optic interactions in fibres and crystals, it is possible to make accurate measurements of these parameters (17). Although the technology is well advanced, there are problems of interference due to vibration, for example, and the question of acceptance by the electricity industry is one which must be addressed. The question of the retraining of personnel involved is one which arises especially with this technology and in particular safety aspects, with high voltage and current must be very carefully considered. Safety standards have been specified for existing technology and the extensive use of new techniques will mean that new standards must be set. The lesson here for the monitoring of environmental parameters is that safety standards must be carefully considered and set, but must not impede the development of new devices.

4. Interferometric Sensor Techniques

A number of different techniques using interferometers as a means to extract accurate distance measurement information have been proposed in the literature. A review of these techniques has been published recently by Jackson (5) and in this approach, the measureand is usually converted to a displacement of one of the mirrors of the device and this is determined by the use of the interferometer. Interferometers are very sensitive to small changes in displacement and to the movement of fringes, corresponding to very small displacements, on the sub-micron region. However one of the difficulties in experiences with these devices is that of the signal ambiguity that arises from the loss of the "zero position" on the light sources switched off and cross-sensitivity i.e. a sensitivity to pressure may also exist when temperatures is being measured.

However FTIR (Fourier Transform Infra Red) spectroscopy incorporating a Michelson interferometer has gained wide acceptance and there is potential for optical versions of these devices.

Need for Chemical Sensors

As an illustration of the need for sensors which detect the presence of chemicals in the environment, the field of gas sensors is one where there has been a considerable upsurge of activity in the last five to ten years. Gas sensors provide a vital interface between automatic and robotic systems and the environment, although the technology still lags that of computing and electronics. The specific areas where gas sensing devices are important are in hazard monitoring, industrial hygiene, pollution monitoring and control, process control, combustion monitoring and control and additionally in medical applications. In considering the need for gas sensors for environmental monitoring, fibre optics can provide several solutions, but there are in addition a number of alternative means of sensing many gases. For example, calorimetric cells, electrochemical cells, semi-conductor devices (inorganic or organic), FETs, or piezo-electric (surface acoustic wave) devices. Thus the analyst who wishes to make a measurement within the environment has a considerable choice as to the most suitable technique to utilize in terms of the sensor he chooses. As a result, if fibre optic sensors are to make a particular impact upon environmental monitoring, then it is important that they can compete with the range of technology that has been described above.

Chemical Sensing using Fibre Optics

Basically the chemical and environmental sensor consists of either a single optical fibre which is connected to the reagent phase at the end of the fibre where the interaction occurs, a bifurcated fibre, where the input and output to the reagent phase may be through an optical coupler, or possibly a third scheme where evanescent wave coupling occurs to the sensor material, where the cladding of the fibre is stripped away and the interaction occurs through material deposited on the fibre itself as shown in Fig. 6. The device however will essentially consist of the following four features:

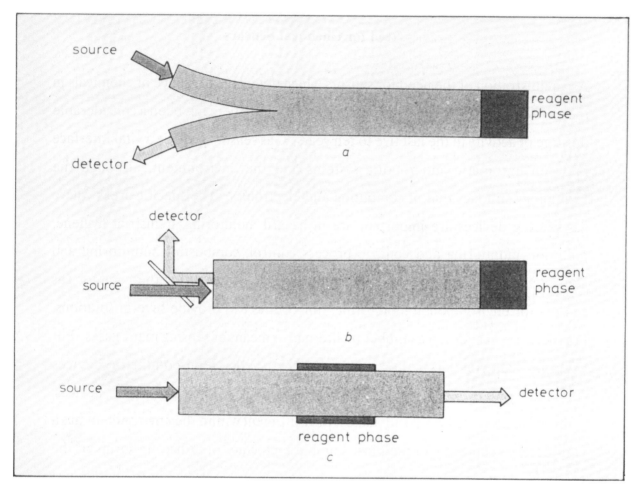

Three types of optical-fibre chemical sensor[1]: (a) Separate fibres carry light from the source to the reagent phase, and the reagent phase to the detector; (b) A single fibre is used for both light paths; (c) The reagent phase is positioned on the outside of an unclad optical fibre, and causes modification of the light transmission

Figure 6 : Illustration of placement of reagent phase in optical fibre sensor devices.

1) A sensing transducer.

2) An optical fibre transmission path.

3) An opto-electronic interface.

4) Signal conditioning/Processing electronic.

In order to consider if fibre optics can meet the task, it is valuable to consider each of these features and the impact that fibre optics has upon them.

1. Sensing Transducer

The sensing transducer must be specific to the species to be measured, or at least specific information must be obtained. It is possible that if a sensor is sensitive to several species that the information on the specific species required can be obtained by, for example ratioing of the output of several sensors or subtraction of a background signal obtained by one sensor from a signal which contains the desired information and the background signal via a second sensor. The reagent must not contaminate the sample (unless the sample is disposable), and this is particularly important where toxic chemicals are used for the reagent phase. It is a specific feature of the biological market that *in vivo* sensors must not damage the patient. In order to avoid contamination, sensors may have to be constructed using suitable membranes which allow the sensing material in but not the reagent phase out, or immobilized reagents which are physically located on a suitable subtrate may be used. One difficulty that arises with these techniques is speed of response, which is generally greatly slowed by the inclusion of such techniques.

2. The Optical Fibre Transmission Path

It is important the correct wavelengths are used in optical fibre sensors. When a non-intensity dependent parameter is measured, there is no need for referencing through the use of a further optical signal, but frequently a second wavelength, which is unaffected by the reagent phase, is used and a ratio is obtained by measuring absorption, for example at two wavelengths. It is important that these wavelengths are adequately transmitted by both the fibre and the reaction cell, and that fibre transmission is suitable for both UV and infrared transmission, if these are used in the sensor device. As an example, many absorptiometric sensors rely upon UV transmission and optical fibres tend to have a much higher loss in the ultra violet. Specialised fibres must be chosen for operation below 250nm and indeed into the vacuum ultra violet region.

3. Opto-Electronic Interface

The sensitivity of detectors must again be suitable to match the wavelength of operation of the sensor technique and there is a preference for the use of solid state detectors to avoid fragile and high voltage photomultiplier tubes. However if signal levels are low these devices may have to be used in spite of the above difficulties.

4. Single Conditioning/Processing Electronics

Progress in recent years in this area has been swift and through the availability of desk top PCs, and the use of algorithms programmed into these or into microprocessors dedicated to the task, correction tables, look-up tables, and other data can be stored to enable the measureand information to be extracted. This is a fundamental requirement for a reproducible and accurate result.

Environmental Monitoring
- Areas that may be addressed by Fibre Optic Sensors.

There is a wide variety of monitoring areas which may be approached through the use of fibre optic sensors and several of these are listed below. For example, in the thermal treatment of solid waste, it is necessary to monitor temperature to determine the way in which the reaction is proceeding and in such applications as odour control in the treatment of waste, gas monitoring sensors are essential. Industrial effluent may be contaminated by the presence of heavy metals which are harmful to health, and so it is necessary to develop sensors to monitor several species of interest. Again in industrial effluent or air emission, it is important to reduce the level of organic compounds which are present and techniques therefore are being developed using fibre optics to monitor these compounds and to determine the change in the level of pollutant present. Of particular topicality is the need for the reduction of chloro-fluorocarbons (CFCs) and halons to the atmosphere and so this again presents a significant monitoring problem for the environmental sensor. In spite of the careful control of the industrial environment, occasionally spillages and contaminations exist and so it is vital to be able to monitor such harmful species when such accidents occur.

As a result there is a need for a wide range of sensing devices for sensing in water and in air, in addition to such basic parameters as solution pH. Many of the techniques that have been applied to the monitoring of physical sensors can be applied also to chemical and environmental sensors. As an illustration, temperature sensing through the use of fluorescence techniques has being particularly effective, as discussed. This arises because it is the monitoring of an non-intensity dependent parameter, solid state sources such as LEDs or diode lasers are employed, solid state detectors are used, as is standard optical fibre i.e. fibre which is available in long length and reasonable cost. Full advantage is taken in such systems of microprocessor based signal averaging techniques and there is overall a lack of cross- sensitivity in the temperature measurement solutions that have been employed. Most importantly, the market has been clearly identified and the sensor fits into that particular niche which is otherwise unfulfilled. Such techniques, for example, have been employed in the monitoring of pH by Wolfbeis et al (18) and Grattan et al (19) using this sort of technology for the monitoring of pH of titration solutions. In addition techniques using immobilised dyes have been described with workers at UMIST (20) and in addition gas sensing for example methane sensing has been undertaken by Samson et al (21) to determine the level of this gas in Australian mines. A simple absorption technique has been employed using an infrared wavelength and very successful results have been obtained, by comparison to those obtained using conventional techniques. Further work at BHP in Australia by Sampson et al (22) again using absorption techniques, has been for the monitoring of Cu II concentrations in salt baths used for chlorine leaching of chalcopyrite. The approach, as illustrated in Fig. 7, is a simple absorption cell which is dipped into the sample and again solid state sources and detectors are employed together with simple signal conditioning. Thus elementary principles of spectroscopy coupled with careful association of the approach used to the spectral features of the material to be sensed have been employed successfully, to produce simple sensors which mimic some of the successful techniques of physical sensors.

Resonant structures have been considered for employment in gas sensing for some time and a recent review by Alder and McCallum (23) has discussed the sensing of

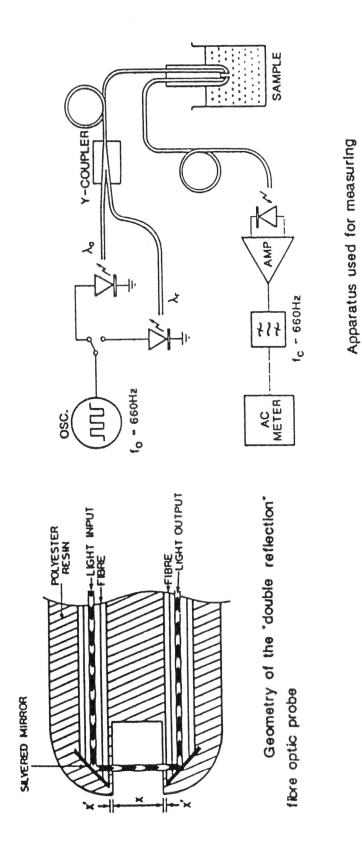

Figure 7 : Absorption sensor for determining of chemical species.

paper concentrated upon the employment of electrical devices to sense the resonant frequency change, but as this can also be detected optically, there is considerable scope for the development of all -optical gas sensors using resonant devices, which would be inherently safe as no currents would flow at the sensor head itself. With an output in frequency form such non-intensity dependent sensing devices would be particularly powerful and the application of coatings to the resonator could give high sensitivity through the monitoring of mass changes through the chemical reactions which may occur in the sensor itself. Quartz or less suitably, silicon components may be used and it may even be possible for these to be disposable in view of the considerable production of resonant elements for other applications, thereby exploiting an existing source of supply. There may be some problems with the chemistry involved, for example, the cross- sensitivity of the device or the difficulty in making a device sensitive enough to monitor at the level of gas experiened, but these are problems which must be addressed in further research. Gaseous species such as ammonia, carbon dioxide, carbon monoxide, cyanides etc. can be detected using this approach.

Interferometric techniques have been used to determine gaseous species, for example on work of Butler et al (24) where a Michelson interferometer was employed together with the palladium coated fibre which was sensitive to hydrogen. This fibre formed part of one arm of the interferometer and as the hydrogen reacted with the palladium coating, the optical path lengths changed the fringe pattern produced in the interferometer. Problems exist however with reversibility and with obtaining specific chemistry sufficiently specific to make such sensitive measurements, but in principle simple and applicable physical sensor techniques have employed for chemical monitoring. Recently DTE in the UK has produced a specification for an optical transform image modulation gas sensing device (25) and this employs again sensitive Michelson interferometric techniques and the change in the fringe pattern produced as a result of the presence of the gas indicates the presence of a particular species. Other illustrations exist in the medical field where a considerable level of sensor requirement exists, largely unmet by current technology. It is here that fibre optic sensors can be developed and it is a field where disposability is generally important (thereby eliminating some problems of sensing at low cost. In addition, the use of "on-line" remote analysis of the environment using fibre optic techniques is an approach that can reduce costs in terms of sample collection and sample preparation, and the automating of the process of data feed-back and data distribution, again can be of considerable advantage in lowering the overall cost of such operation, with savings of manpower and resources.

Can They Meet The Task ?

In looking at the answer to this question, several points have been raised in the course of this paper. The considerable experience in physical sensing that already exists must be exploited and the lessons that have been learned from the development of these devices should be employed in the development of fibre optic environmental sensors. There are some aspects of chemical sensors that already have been well developed for laboratory use, and there is a particular need to ensure that these can be ruggedized and employed outside the laboratory environment. In addition it is important that these are as compatable as possible with modern data transfer techniques. There will be a need to retrain staff to use the new technology, but as this is happening already in the development of new communication systems, this may not be as major a problem as had been envisaged previously. Signal processing techniques that are offered through the use of microcomputers and microprocessors can allow relatively unsophisticated sensing mechanisms to be used with corrections obtained in the signal processing and data and look-up tables stored in the electronics.

Fibre optic sensors offer unique advantages, and these advantages must be exploited in their use and environmental monitoring. There is potentially a very large industrial and medical market which can provide a considerable level of support for developments in the field. In addition the possibility of hybrid optical and optical with electronic or optical and electrochemical devices must be considered, and this may well provide the most satisfactory solution to some sensor problems.

The sensing of the environment and its careful monitoring and control is particularly important as the twenty first century approaches. Fibre optic sensors have a vital role to play in the development of new sensing systems to meet this task. On the basis of the experience gained with the development of fibre optic sensors for physical measurements, it is expected that there will be devices available able to meet this need and commercial products will be produced in some quantity over the next several years to begin to address the sensing problem.

References

1. N.P. Ludlam

   "Safety of Optical Systems in Flammable Atomspheres"

   Confidential report to UK OSCA, number 55/555 (unpublished), 1988.

2. J.P. Dakin

   "Fibre Optic Sensors - Principles and Applications"

   Control and Instrumentation p41, September 1984.

3. Accufiber Corp.

   Technical Literature, Vancouver, 1983.

4. A. Dandridge, A.N. Tveten, G.H. Sigil, E.J. West, T.G. Gaillorenzi

   "Optical Fiber Magnetic Sensors"

   Electron. Letts, 6, 636, 1980.

5. D.A. Jackson

   "Monomode Optical Fibre Interferometers for Precision Measurement"

   J. Phys. E: Sci. Instrum. 18, 981, 1985.

6. B.T. Meggitt, A.W. Palmer, K.T.V. Grattan

   "Fibre Optic Sensors using Coherence Properties - Signal Processing Aspects"

   Int. J. Optoelectronics 3, 451, 1988.

7. York Technology

   DTS System II Manufacturer's Publicity Data, 1988.

8. A.P. Appleyard, P.L. Schrivener, P.D. Maton

   "Intrinsic Optical Fibre Temperature Sensor based on Differential Absorption Techniques"

   Rev. Sci. Instrum. - to be published 1990.

9. G.D. Pitt, P. Extance, R.C. Neut, D.W. Batchelder, R.E. Jones, J.A. Barnett, R.H. Pratt

   "Optical-Fibre Sensors"

   IEE Proc. 132J, 214. 1985.

10. K.T.V. Grattan

"Fibre Optic Techniques for Temperature Sensing - a Review"

In Fibre optic sensors for chemical and biosensors Ed O.S. Wolfbeis, Fl, USA, CRC Press - 1990.

11. J.P. Dakin

"Temperature Measurement using Intrinsic Optical Fibre Sensors"

Int. J. Opt. Sens. 1, 101, 1986.

12. K.A. Wickersheim and R.V. Alves

Biomedical Thermology p547 Pub: Alan Liss N.Y. USA, 1982.

13. K.A. James, W.H. Quick

"Digital Fibre Optic Sensors"

Control Engineering 30 (Feb 1979).

14. K.T.V. Grattan, A.W. Palmer

"Fibre-Optic-Addressed Temperature Transducers using Solid-State Fluorescent Materials"

Sensors and Actuators 12, 375, 1987.

15. K.A. Wickersheim

"Commercial Applications of Fiber Optic Temperature Measurement"

Proc. 'Fiber Optic Sensors IV', the Hague, the Netherlands, March, 1990. Proc SPIE - to be published 1990.

16. R.C. Spooncer, B.E. Jones, G.S. Philp

"Hybrid and Resonant Sensors and Systems with Optical Fibre Limits"

J. IERE 58, 585, 1988.

17. A.J. Rogers

"Optical Methods for Measurement of Voltage and current at High Voltage"

Opt. and Laser. Tech. p273, 1977.

18. O.S. Wolfbeis, B.P.H. Schaffar, E. Kaschnitz

"Construction and Performance of a Fluorimetric Acid-Base Titrator with a Blue LED as a Light Source"

Analyst 111, 1331, 1986.

19. K.T.V. Grattan, Z. Mouaziz, A.W. Palmer
    "Dual Wavelength Optical fibre Sensor for pH Measurement"
    Biosensors 3, 17, 1987.

20. R. Narayanaswamy, F. de Sevilla
    "Optical Fibre Sensors for Chemical Species"
    J. Phys. E. Sci. Instrum. 21, 10, 1988.

21. P.J. Samson, A.D. Stuart
    "Coal Mine Methane Sensing by Optical Fibre"
    (IREE, Aust.) Tasmania, Dec. 1988, p113.

22. P.J. Samson
    "Simple Double Reflection Fibre Optic Probe for On-Line Spectroscopy"
    Proc. 11th Aust. Conf. on Optical Fibre Technology (IREE, Aust.) Geelong, Dec. 1986, p197.

23. J.F. Alder, J.J. McCallum
    "Piezoelectric Crystal for Mass and Chemical Measurement"
    Analyst 108, 1169, 1983.

24. M.A. Butler
    "Optical Fiber Hydrogen Sensor"
    App. Phys. Letts. 45, 1007, 1984.

25. DTE Enterprises, Malvern, U.K.
    "OTIM Passive Remote Gas Detector"
    Unpublished data, 1990.

MONITORING BEYOND THE BOUNDARY FENCE

DR. B. CALLAN
PENN CHEMICALS B.V.
CURRABINNY, CARRIGALINE, COUNTY CORK, IRELAND

ABSTRACT:

Air pollution has traditionally been measured in terms of ambient sulphur dioxide and smoke concentrations. However, more appropriate measurements in the vicinity of the chemical industry are centred on the sampling and analysis of organic vapours, particularly in the case of reduced sulphur compounds. The latter group of chemicals, such as methyl mercaptan and hydrogen sulphide have very low odour thresholds and consequently pose difficult problems for both sampling and analysis. Measurement strategies and philosophies are outlined with particular attention centred on odour detection and measurement using odour patrols, the scentometer, the olfactometer and the butanol intensity scale.

INTRODUCTION

Air quality in Ireland is generally measured in terms of the sulphur dioxide and smoke concentrations present, with additional measurements of ambient lead in Dublin and Cork and ambient nitrogen oxide measurements in Dublin (1). The method normally used for sulphur dioxide and smoke measurements is the British standard BS 1747 (2), where air is drawn sequentially through a Whatman filter paper followed by a dilute peroxide solution. The smoke stains the filter paper black and is subsequently measured by reflectance while the sulphur dioxide is oxidised to sulphuric acid in the peroxide solution and subsequently measured by titration. Although this method is prone to inaccuracies of over or under estimation of the true sulphur dioxide concentrations (3) it nonetheless provides a useful index of ambient air quality and has been used to verify same in the vicinity of both major powerplants, such as the 900 MW coal fired unit at Moneypoint on the west coast of Ireland (4) and in the vicinity of major chemical plants, such as Merck Sharp and Dohme at Ballydine, Co. Tipperary.

This method when used to monitor industry needs to be supplemented with more specific measurement of gaseous pollutants, particularly those with low odour thresholds. The latter is the case with reduced sulphur compounds, such as mercaptans and hydrogen sulphide, which are used or produced by several of the chemical industries in the Cork area. A comparison of some typical odour thresholds, with their respective irritation concentrations and work place threshold limit values (TLV's) is presented in Table 1 (5,6).

TABLE 1

COMPARISON OF CHEMICAL ODOUR THRESHOLDS AND IRRITATION CONCENTRATIONS

CHEMICAL COMPOUND	LOW ODOUR mg/m^3	HIGH ODOUR mg/m^3	IRRITATION CONCENTRATION mg/m^3	TLV mg/m^3
Carbon Tetrachloride	300	1500	-	30
Chloroform	250	1000	20480	50
Acetone	47	1614	475	1780
Isopropyl Alcohol	7.4	490	490	980
Acetic Acid	2.5	250	25	25
Formaldehyde	1.5	74	1.5	1.5
Sulphur Dioxide	1.2	13	5.0	5
Methyl Ethyl Ketone	.74	148	590	590
Ammonia	.027	40	72	18
Carbon Disulphide	.024	23	-	30
Dimethyl Sulphide	.0025	.0580	-	14
Hydrogen Sulphide	.0007	.0140	14	14
Dimethyl Disulphide	.0001	.347	-	
Methyl Mercaptan	.00004	.082	-	1.0

From this table it can be seen that methyl mercaptan is 30 thousand times more odorous than sulphur dioxide and 7.5 million times more odorous than carbon tetrachloride, the characteristic smelling solvent used in dry cleaning. The normal odour threshold quoted for this compound is 0.6-4.0 $\mu g/m^3$ (7-10), with $2\mu g/m^3$ most frequently used. Thus, using a value of $2\mu g/m^3$, this compound can be detected 500 times lower than its Threshold Limit Value (TLV), the concentration at which nearly all workers (40 hour week) can be repeatedly exposed, day after day, without adverse effect (6). Thus, the problem with this pollutant is one of odour (nuisance) rather than adverse health effects and in the vicinity of chemical plants which use or produce such compounds it is necessary to detect them at extremely low levels.

The techniques used to measure these compounds are numerous and in principle include all the techniques of analytical chemistry. The very low concentrations of these reduced sulphur compounds are frequently below the detection limits of available instrumentation making accurate real time measurements impossible.

Within Penn Chemicals B.V. a dedicated process gas chromatograph (Bendix), sampling up to ten plant locations, is used to continuously monitor ambient levels of methyl mercaptan and dimethyl disulphide. This instrument is designed to alarm in the event of designated levels of the contaminant being detected. However, while making an excellent alarm instrument it does not have the sensitivity to monitor the low $\mu g/m^3$ concentration at which these compounds are odorous and subsequently would be of little value beyond the boundary fence. On the other hand, this gas chromatograph can accurately measure concentrations of above $40\mu g/m^3$ of these contaminants, a concentration which if measured on site would likely be diluted below the odour threshold of 2ug/m^3 by the time it crossed the boundary fence. Similarly a dedicated quadruple mass spectrometer (VG Petra System) is used to continuously monitor the plant environment for potential contaminants such as carbon disulphide and dimethyl sulphate. The latter instrument may not be used for the monitoring of low molecular weight reduced sulphur gases such as mercaptan and hydrogen sulphide because of the many ambient interferences at these low masses.

Thus most methods used beyond the boundary fence involve concentrating the pollutants by trapping them on adsorbents or in absorbing solutions for later analysis. This, however, merely confirms that the compound was present as a historical event. An alternative strategy, as used at Penn Chemicals BV is to evaluate the odour directly using trained personnel and internationally recognised instrumentation and to combine this with chemical analysis of grab samples. The former method (ie concentration on adsorbents) is then used as an evaluation of the general ambient air quality while the latter may be used as an evaluation of ambient air at any given time and subsequently as a control mechanism if necessary.

CONCENTRATION OF LOW LEVEL AMBIENT POLLUTANTS

Gaseous pollutants are most commonly sampled using solid adsorption tubes or liquid absorbing solutions, with the former method used most in the detection of reduced sulphur gases. The two most widely used adsorbents are charcoal and tenax with the former being the basis of most of the NIOSH methods of trapping gaseous contaminants (11) and the latter being used more widely with the rapidly growing applications of thermal desorption (12, 13). Other adsorption tubes include activated aluminas, silica gel, molecular sieve, celite and several chromatographic support coated packings such as the Chromosorb and Poropak series. A further form of sample concentration which is often used with an adsorbent is cryogenic sampling. This procedure uses a sampling trap at subambient temperatures causing an ambient vapour to condense and be retained (14). This latter technique does not lend itself to field use and also tends to collect large quantities of water vapour, presenting a major problem in subsequent chromatographic analysis (13 - 14).

Using an adsorption tube it is important to accurately measure the volume of air sampled with a suitable gas meter or using pumps which have been calibrated with the sampling device inline. It is also necessary to demonstrate that the adsorption tube chosen can effectively trap the required pollutant and that once trapped it can be effectively removed for analysis (adsorption/desorption efficiencies). This is best done by injecting a known amount of the pollutant, in the gaseous phase, into a tedlar bag and then sampling using the selected adsorption tube. Alternatively the compound may be injected directly on to the adsorption tube and a volume of clean air, approximately the same as that to be sampled in the field, drawn through the tube. The adsorption tube may then be desorbed, preferably using thermal desorption or alternatively using solvents. Most thermal desorption units use a two stage process where the adsorption tube is first heated in a stream of inert gas in order to desorb any volatile organic compounds present. These are then passed to a cold trap, cooled by liquid carbon dioxide, liquid nitrogen or electrically. This trap concentrates the volatile compounds which are subsequently released into the gas chromatograph in a sharp 'plug' by a sudden increase in temperature. However, using solvent desorption, carbon disulphide is the preferred solvent (using Flame Ionization Detection) for the measurement of most of the common organic solvents such as acetone and isopropyl alcohol, while for the reduced sulphur gases, acetonitrile or toluene are more appropriate. Particular care must be taken using solvent desorption as gases may be desorbed in a different form to their initial presence in the ambient atmosphere, eg. methyl mercaptan when trapped on charcoal is desorbed as dimethyl sulphide when using acetonitrile as the solvent. Certain other adsorbents have been reported to cause related isomerisations and so alter the sample (15,16). Other possible interferences in this form of sample collection include the deactivation of the adsorbent by water vapour or the reaction of the trapped gases with each other on the adsorbent surface. Most commercial adsorbents contain two sections, the second being a back up. Both must be analysed routinely to ensure that the initial adsorbent section has not been overloaded giving a break through of pollutant into the back up section. Should a break through of more than 25% occur it is likely that there will be a loss of sample (11). Break through can also occur if the sampling flow is too great.

In August 1987, Penn Chemicals B.V. located three ambient monitors in the vicinity of its plant in order to monitor ambient organic pollutants. These monitors have been independently operated on behalf of Penn by EOLAS, with independent sample collection and analysis. The monitors are capable of collecting eight discrete 24 hour samples on consecutive days (approximately $1m^3$/day) followed by weekly collection and analysis. The initial adsorbent used was charcoal, followed by thermal desorption and gas chromatographic analysis for methyl mercaptan using a photo ionization detector (see below).

However, the charcoal has been replaced with tenax and since October 1989 each of the samples have received a full gas chromatographic analysis with mass spectroscopy detection (GC/MS). Only two of the monitors are located in the direct vicinity of Penn Chemicals B.V. but not in the direct vicinity of any other industries and these have shown only a very low level of organic contaminants. No reduced sulphur compounds have been detected at either of these monitors since GC/MS analysis began six months ago with only isolated detection of organic solvents such as isopropyl alcohol and methyl ethyl ketone at concentrations less than $0.5 \mu g/m^3$ being measured.

A wide range of individual gaseous components may also be monitored using absorption solutions, similar to that described above for sulphur dioxide (2), with absorbents such as cadmium hydroxide for H_2S (17) and an aqueous solution of mercuric acetate-acetic acid for methyl mercaptan (18). Alternatively, chemically impregnated filters may be used such as the commonly used lead acetate for H_2S (19, 20) and silver nitrate, mercuric cyanide and mercuric nitrate - tartic acid for a range of reduced sulphur gases such as methyl mercaptan, dimethyl sulphide, dimethyldisulphide and hydrogen sulphide (21,22). The advantage of these methods is that less sophisticated equipment is needed but they are generally more prone to chemical interference and suffer the same drawbacks as adsorbent sampling in that large volumes need to be sampled in order to reach the required detection limits for the compounds under investigation.

Thus the major disadvantage of the techniques just described is the inability to produce real time measurements. Although such instruments exist and operate well for sulphur dioxide (based on fluorescence) and nitrogen oxides (based on chemiluminescence) the same is not true for the reduced sulphur compounds (1). Recent reports have indicated a universal sulphur detector, based on a chemiluminescence technique, that would be capable of such real time measurements, but this is as yet in the embryonic stage of development (23). The inability to give real time measurements means that low level contaminant emissions, which may be odorous but lasting only minutes, are not detected by these forms of measurements as the short duration emission is effectively diluted by a much larger clean air sample.

GRAB SAMPLING OF LOW LEVEL AMBIENT POLLUTANTS

The closest approach to real time sampling is to take very short measurements using adsorbents or absorbents (e.g. 10 - 30 minutes) or to take grab samples which effectively give a snapshot of the contaminant profile at the time of sampling. Grab samples are best taken using one of the commercially available gas sampling bulbs. These contain two teflon stopcocks and an injection port containing a PTFE coated septum. The two stopcocks are necessary for the effective flushing of the bulb prior to sampling in order to avoid cross contamination of samples. The stopcocks are then used to evacuate the bulb and although the vacuum will remain for several hours, the bulb is best evacuated immediately prior to use. The bulb is then used to take a sample by opening one of the stopcocks when an odour is detected and thus drawing in a representative sample. Alternatively a sample may be drawn into an inert tedlar bag using an airtight container with only the inlet to the bag exposed to the atmosphere. The air surrounding the bag is then removed using a pump, causing the bag to inflate (Fig. 1). The tedlar bag may also be filled from the discharge side of the vacuum pump if the exposed parts of the pump are made entirely from an inert material such as PTFE. Pumps with stainless steel components which come into contact with the sample may not be used as stainless steel adsorbs many of the reduced sulphur compounds (24). In practise the sampling of reduced sulphur compounds in glass bulbs is preferred, as low level concentrations of these gases are more stable in glass. As a general rule low level concentrations of reduced sulphur compounds must be analysed within an hour when using tedlar bags, to maintain sample integrity.

These bags are also much more difficult to clean as they must be flushed and evacuated repeatedly. This is done at Penn Chemicals using a device specifically engineered (in-house) for this purpose, where five bags are filled to capacity with nitrogen and subsequently evacuated in three minute cycles. However, a major advantage of the tedlar bag is the ability to further analyses the sample using a large selection of colorimetric detector tubes (25, 26) or using an olfactometer to determine the dilution to odour threshold.

CHROMATOGRAPHIC ANALYSIS

The grab samples are then analysed for a range of reduced sulphur compounds using gas chromatography (GC). A range of GC columns may be used for this analysis, namely Chromosil 310 and 330 (silica gel based) Carbopack B (deactivated graphitised carbon base), Chromosorb T

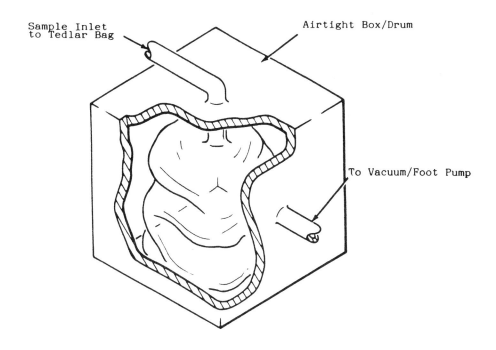

FIG. 1   TAKING SAMPLES IN TEDLAR BAGS

(polyphenyl ether base), and Poropak QS (27, 28). The latter two chromatographic columns are routinely used at Penn Chemicals, both columns being packed in teflon, Fig. 2. A large sample volume, 15ml, is introduced using a sampling loop. The components are then chromatographically separated and detected first using a photo ionization detector (PID) followed by a flame ionization detector (FID), as depicted in Fig. 3.

The PID contains a sealed source which emits photons in the far ultra violet which are sufficiently energetic to ionize molecules passing through the ionization chamber provided they have an ionization potential less than that of the lamp used (10.2V emission line of hydrogen used). The ions are driven to a collector electrode by an electrical field with the current being amplified and measured (10,29-31). As the PID is non destructive the sample can then be passed directly to the more common FID which, as its name implies, involves the ionization of organic components in a flame. A hydrogen - air flame is used because of the low concentration of ions. The introduction of organic carbon compounds into the flame results in the formation of ions which reduce the electrical resistance of the flame. This change in resistance is measured by two electrodes adjacent to the flame to give the component response.

The other widely used sulphur detector is the flame photometric detector (FPD), also used at Penn Chemicals in a process GC dedicated to continuous plant ambient air analysis. In this detector the sulphur containing species is burned in a hydrogen rich flame and the sulphur (S) converted to an excited S_2 at a higher energy state.

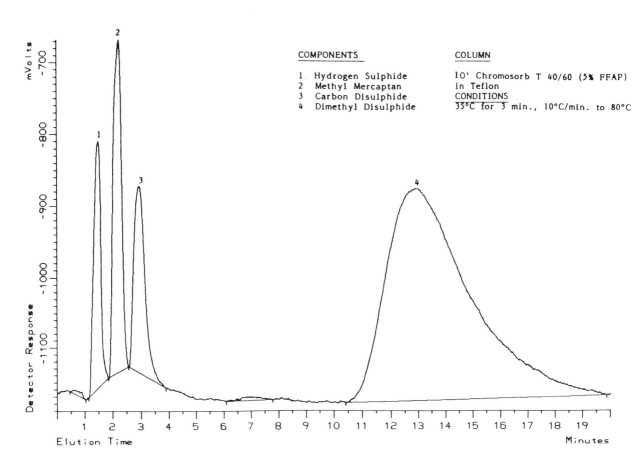

FIG. 2   PID DETECTION OF H_2S (2780ug/m^3), MMC (785ug/m^3), CS_2 (750ug/m^3) AND DMDS (1540ug/m^3).

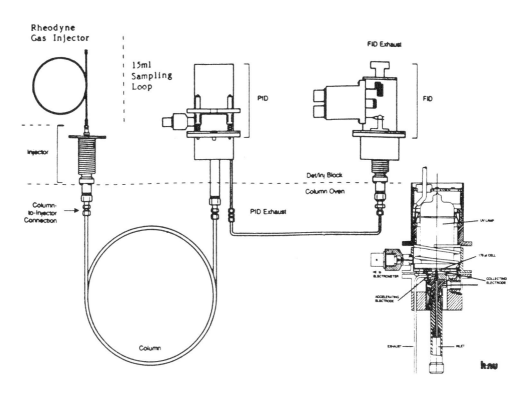

FIG 3   GAS CHROMATOGRAPH SCHEMATIC AND PHOTO IONIZATION DETECTOR (BOTH HNU SYSTEMS LTD.)

This S_2 returns to the ground state with an emission of light of characteristic wavelengths (384 and 394 nm) which are subsequently used for the quantification of the sulphur containing species (32).

The advantage of a dual detection system, PID followed by FID, is that the relative responses of both detectors give an indication of the type of compound present, eg. reduced sulphur compounds which have a good PID response have little or no FID response. The PID, although not a sulphur specific detector like the FPD, has been found to be more sensitive as well as having a wider dynamic range of measurement. While both these detectors are more sensitive than GC/MS they cannot replace MS for unequivocal component identification. However, MS may not be used with the large sample volumes as they are not designed for capillary chromatography unless some form of cryogenic concentration is first used on the air sample. In the latter case an OV 101 capillary column may be used for the reduced sulphur compounds. Although detection limits of 2-4 $\mu g/m^3$ have been reported for some mercaptans using a 5ml sampling loop (10) the best detection limit of the PID is 20$\mu g/m^3$ for methyl mercaptan and dimethyl disulphide with the system described above. As this value is ten times greater than the odour threshold, the chromatographic analysis is only of use to confirm a present odour at this level and to identify the odorous components. It does also provide analysis of any other high concentration organic components which may be non odorous. Thus, the classical techniques of analytical chemistry have only a limited use when dealing with the low odour threshold species such as the reduced sulphur compounds.

ODOUR MEASUREMENT

The direct measurement of odour is a method well suited for the monitoring of industry which uses reduced sulphur compounds. The German TA Luft (33) defines the unit of odour as the olfactometrically measured ratio of volume flows necessary to dilute a sample gas down to its odour threshold (ie. the number of times an odorous sample must be diluted with an equal volume of odour free air to reach the lowest limit at which the odour is detectable), commonly referred to as the dilution-to-threshold (D/T). This form of measurement has been common in the US over the past two decades and although no odour control regulations exist at Federal level (ie. EPA regulations) many regulations do exist at both state and local levels. A summary of some typical regulations are outlined in Table 2, with the D/T values measured using a scentometer or an olfactometer. In general, persistent odour above 7 D/T will cause complaints while those above 31 D/T are considered a serious nuisance (34). The official American Society for Testing Material (ASTM) procedures (35 - 37) are widely used.

Odour is a physiological individual response and only when this response can be replicated can odour measurement be placed on an objective basis. The most reliable and accurate instrument of measuring odour is the human nose (9) and with this in mind Penn Chemicals B.V. use a system of odour patrols where for 18 out of every 24 hours, trained personnel are located off site down wind of the plant. The patroller is in radio contact with designated personnel on site and in direct telephone contact with others when needed.

TABLE 2

US LEGISLATION BASED ON D/T VALUES

(MEASURED ON SCENTOMETER/OLFACTOMETER)

LOCATION	DATE	D/T RESIDENTIAL AREA	D/T COMMERICAL AREA
Colorado	1971	7	15
Missouri	1970 Rev. 1984	7	20
North Dakota	Rev. 1987	2	
San Francisco Bay Area California	Rev. 1982	2	
Wyoming	1970	7	

Should an odorous emission occur from the plant, it may be quickly reported and dealt with accordingly. Each of the odour patrol personnel have been professionally screened by odour consultants (38) using recognised methods (35 - 44).

Potential odour patrol personnel and odour panelists are first screened for their ability to detect n-Butanol vapours above aqueous Butanol solutions in odour free containers. Individuals are first presented with three containers, two containing blank water and the third a sub-threshold n-Butanol solution. A series of similar triangular presentations are made, with the Butanol content of one of the containers doubling each time, and the individual is forced to choose one of the containers even if unable to detect any odour. The concentration at which the individual correctly identifies the n-Butanol odour and continues to identify it correctly is that individual's odour threshold for n-Butanol (36, 37, 42). Individuals are then required to compare odour intensities of prepared 'unknowns' with the recognised Butanol intensity scale (Table 3) and in order to proceed with the odour screening each individual must match intensity of the sample within one scale unit (37,39). This n-Butanol scale, representing odours from sub-threshold (intensity 1) to very strong (intensity 8) is subsequently used on odour patrols. Odour levels below intensity 3 have been found to be acceptable to 80% of a population, with intensities 4 - 6 causing a probable nuisance and intensities above 6 causing a definite nuisance (38). The n-Butanol series is generally memorised before going on an odour patrol and only referred to as needed to avoid olfactory fatigue. The only new US odour regulation in the past decade was made by the state of Louisiana, adopting this n-Butanol scale as the basis of ambient odour

TABLE 3

BUTANOL INTENSITY SCALE

INTENSITY	AIR ppm	LIQUID ppm	LIQUID PREPARATION
8	2000	20000	8ml of BuOH in 400ml H_2O
7	1000	10000	200ml of 8 plus 200ml H_2O
6	500	5000	200ml of 7 plus 200ml H_2O
5	250	2500	200ml of 6 plus 200ml H_2O
4	120	1200	200ml of 5 plus 200ml H_2O
3	60	600	200ml of 4 plus 200ml H_2O
2	30	300	200ml of 3 plus 200ml H_2O
1	15	150	200ml of 2 plus 200ml H_2O

intensity measurement (38). Butanol is used in these odour intensity measurements as it is considered to have a neutral odour, not pleasant and yet not unpleasant. However, it is replaced by pyridine in Sweden and France using the argument that it is necessary to have something unpleasant.

Each individual's sensory perception is further evaluated using a dynamic dilution olfactometer initially with inflated tedlar bags containing known n-Butanol vapours. The olfactometer is used to determine the dilution to threshold (D/T) of samples. The tedlar bag is attached to the olfactometer and the 'odorous' bag sample pulled into the olfactometer by the flow of carbon filtered dilution air flowing through a venturi. The venturi also acts as a mixing chamber for the odorous sample and the dilution air. The flow of odour free dilution air is maintained at a constant flow while the sample flow is adjusted by a high precision rotometer to create the required sample dilutions (35, 42). The olfactometer also uses a forced choice triangular presentation where the diluted sample flows to one of three cone shaped borosilicate glass sniffing ports and carbon-filtered odour free air goes to the other two ports. The flow rate of the ports is normally maintained at 3 litres per minute (normal breathing rate) and each panelist is again required to correctly identify the Butanol odour. The forced choice procedure (Fig. 4) requires the participant to make a choice, indicating if he has got an odour, something which was different but not odorous or whether he is simply guessing. To avoid odour fatigue, the evaluation begins with extremely low concentration of odorous sample (a high dilution) and the threshold is approached from the bottom of the scale rather than the top. The sniffing ports are randomly repositioned

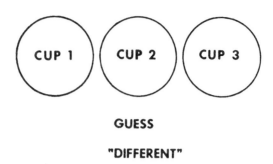

FIG. 4   TRIANGULAR FORCED CHOICE PROCEDURE

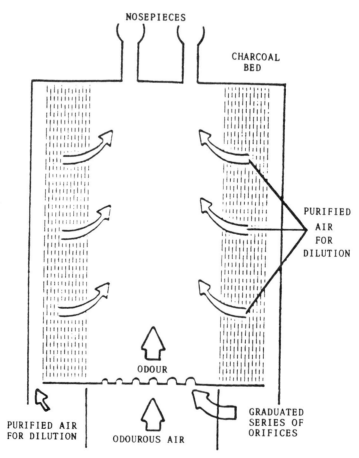

FIG. 5   SCENTOMETER SCHEMATIC

with each change of dilution. Odour quantification consists of
determining the D/T ratio at which 50% of the panel are able to
correctly identify the odorous sniffing port as that containing an
"odour" (35, 36, 39, 43, 44). Odour detection training on the
olfactometer is completed using real samples, trapped in tedlar bags
(Fig. 1) or using source samples, taken at emission points. Thus when
an odour perists while personnel are on odour patrol, the ambient air
is sampled in a tedlar bag and the D/T measured (back on site) using
the olfactometer and a "screened" panel whose ability to detect odour
has been confirmed. The olfactometer is also extremely useful for
measuring the D/T of emission sources and has been used at Penn
Chemicals B.V. in conjunction with an EPA approved computer "puff"
model, much the same way as one does for the dispersion modelling of
conventional contaminants. This model (38) is used to evaluate the
potential for down wind odour impact beyond the boundary fence and
uses the D/T values along with the more conventional physical
characteristics of the emission source. The major feature of the
odour model is the short averaging time of 2 seconds, which is
necessary to predict instantaneous odour concentrations. All
significant emission sources have been evaluated at Penn using D/T
values measured by panels trained as above.

The final training of odour patrol personnel is in the use of the
scentometer, the most common instrument used for D/T
measurements (34). This instrument is particularly useful in that it
gives an on-the-spot odour evaluation. A schematic of the scentometer
is shown in Fig. 5. Odorous air is drawn by inhalation through one of
six graduated orifices and is mixed with air from the same source
which has been purified by passing through two activated charcoal
chambers. Dilution rates are fixed by orifice selection giving D/T

values of 2 to 350 (Table 4). One useful feature of this device is that the user may, by breathing only air purified by the activated carbon (ie. all orifices closed) combat the olfactory fatigue which can occur when present in an odorous area and thus return to zero reference before trying any additional dilution to evaluate the D/T. This value is then combined with the n-Butanol intensity scale evaluation. A critical review of the use of odour measurement techniques applied to the regulations for the control of odours (45) has stressed the importance of both D/T and intensity measurements being made.

Finally, electronic odour monitors also exist (46) using principles such as the sensitive metal oxide thermal conductivity sensor. This instrument is based on temperature changes which result from the molecular gases oxidising when they come into contact with a heated sensor. However, this instrument like several others, which are based on physical concentration measurements, is unlikely to be useful in the monitoring of low concentration reduced sulphur gases. Where ambient contaminants are identified by gas chromatography using a non-destructive detector such as the PID, the detector outlet (waste vent) may be directed to a 'sniffing port' where the separated individual components are further characterised in terms of their odour, this characterisation again being achieved with the human nose. Alternatively, in the case of FID and FPD detectors, the sample may be split between the detector and a sniffing port but this arrangement inevitably leads to a decrease in the sensitivity of the chromatographic detection. This method of measurement serves only to indicate which of the contaminants are odorous, what type of odour they have (eg. rotten cabbage) but is not a measure of the amount of odour such as achieved with the olfactometer (47).

TABLE 4

SCENTOMETER DILUTIONS TO THRESHOLD (D/T)

	ODOROUS AIR INLETS (GRADUATED IN INCHES)					
DILUTIONS TO THRESHOLD	1/2	1/4	3/16	1/8	1/16	1/32
350	-	-	-	-	-	Open
170	-	-	-	-	Open	-
31	-	-	-	Open	-	-
15	-	-	Open	-	-	-
7	-	Open	-	-	-	-
2	Open	-	-	-	-	-

– indicates orifice is closed

CONCLUSION

Air quality in the vicinity of a chemical plant may be effectively demonstrated using a variety of classical analytical techniques. However, where very odorous compounds are used or produced, they may be detected by the human nose even if less odorous compounds are present at much higher concentrations. This high sensitivity and discrimination makes the nose far superior to any available instrumentation for the evaluation of odours, which are a nuisance rather than a toxic danger. This sensitivity may be harnessed through the use of instruments such as the olfactometer where a panel of people are used to evaluate the odour. In situations where a class of compounds form the dominant malodour (eg. reduced sulphur gases) a suitable test such as gas chromatographic measurement is most useful when used in conjunction with odour measurements. While gas chromatographic techniques have detection limits above those at which some compounds are odorous, they are nonetheless an excellent measure of general ambient air quality, particularly of the higher concentration non-odorous compounds.

REFERENCES:

1. Callan, B. and Flanagan, P., "Air Pollution Data-Acquisition and Validation in Ireland", Proceedings of the International Workshop on Harmonisation of the Technical Implementation of the EC Air Quality Directives on Sulphur Dioxide, Lead and Nitrogen Oxides, page 117, Lyon, Nov. 22 - 24, 1988

2. British Standard Institute, "Methods for the Measurement of Air Pollution", BS 1747, Parts 2 & 3, 1969.

3. Callan, B., "Accuracy of the Acidimetric Measurement of Sulphur Dioxide", Irish Chemical News, Journal of the Institute of Chemistry of Ireland, page 22, Winter 1989.

4. Callan, B., McCarthy, F. and O'Connor, E., "Electricity Supply Board Monitoring Assessed", Technology Ireland, 20 (9), 37, 1989.

5. Ruth, J. "Odour Thresholds and Irritation Levels of Several Chemical Substances: A Review," Amer. Industrial Hygiene Assoc. J., 47, A 142, 1986.

6. "Threshold Limit Values and Biological Exposure Indices for 1988 - 1989", Amer. Conference of Governmental Industrial Hygienists, Cincinnati, Ohio, 1988.

7. Leonardos, G., "Research on Chemical Odours, part 1." Arthur D. Little, Inc., Cambridge, Massachusetts, 1968.

8. Adams, D., "Sulphur Components and their Measurement", Air Pollution, 3rd. Ed. page 23, Vol. 3, Academic Press, New York, 1976. (Ed. A.C. Stern).

9. Phelps, A., "Odour and its Measurement", Air Pollution, 3rd Ed., page 335, Vol. 3, Academic Press, New York, 1976. (Ed. A.C. Stein)

10. Stein, V. and Narang, R., Anal. Chem., 54, 991, 1982.

11. NIOSH Manual of Analytical Methods, 2nd Ed., Vol. 1-6, National Institute for Occupational Health and Safety, US Dept. of Health, Education and Welfare, Cincinnati, Ohio, 1977.

12. Ryks, J., Drozd, J., and Novak, J., J. Chrom, 186, 167, 1979.

13. Brown, R, and Purnell, C., "Collection and Analysis of Trace Organic Vapour Pollutants in Ambient Atmospheres", J. Chrom., 178, 79, 1979.

14. Axelrod, H. and Lodge, J., "Sampling and Calibration of Gaseous Pollutants," Air Pollution, 3rd. Ed., page 156, Vol. 3, Academic Press, New York, 1976 (Ed. A.C. Stern).

15. Turk, A., Morrow, J. and Kaplan, B., Anal. Chem., 34, 561, 1962.

16. West, P., Sen, B. and Gibson, N., Anal. Chem., 30, 1390, 1958.

17. Intersociety Committee for Manual Methods of Air Sampling and Analysis, "Methods of Air Sampling and Analysis", 2nd. Ed., Amer. Public Health Assoc., Washington, DC, 1977.

18. Moore, H., Helwig, H. and Grawl, R., Industrial Hygiene J., 21, 466, 1960.

19. Hochheser, S. and Elfers, L., Environ. Sci. Tech., 4, 672, 1970.

20. Natush, D., Klonis, H., Axelrod, H., Teck, R. and Lodge, J., Anal. Chem., 44, 2067, 1972.

21. Huygen, C., Anal. Chem. Acta, 28, 349, 1963.

22. Harrison, R., "Nitrogen and Sulphur Compounds", Handbook of Air Pollution Analysis, 2nd. Ed., page 310, Chapman and Hill, London and New York, 1986 (Ed. R.M. Harrison and R. Perry).

23. Benner, R. and Stedman, D., "Universal Sulphur Detection by Chemiluminescence", Anal. Chem., 61, 1268, 1989.

24. Stevens, R., Mulik, J., O'Keeffe, A. and Knost, K., Anal. Chem., 43, 827, 1971.

25. British Standards Institute, "Gas Detector Tubes", BS 5343, 1976.

26. Leichnitz, K., "Air Investigation and Technical Gas Analysis with Drager Tubes", Detector Tube Handbook, 6th Ed., 1985.

27. Mindrup, R., "The Analysis of Gases and Light Hydrocarbons by Gas Chromatography", J. Chrom. Sci., 16, 380, 1978.

28. Analysis of Sulphur Gases, Supelco GC Bulletin 722K, 1988.

29. Driscoll, J., "Review of PID Application in Industrial Hygiene", Industrial Hygiene News, 1, 3, 1988.

30. Langhorst, M., J. Chrom. Sci., 19, 98, 1981.

31. Verner, P., J. Chrom., 300, 249, 1985.

32. Farwell, S. and Barinaga, C., J. Chrom. Sci., 24, 483, 1986.

33. Technical Instructions in Air Quality Control, TA Luft, Federal Ministry for the Environment, Bonn, 1986.

34. Huey, N., Broering, L., Jutze, G. and Gruber, C., "Odour Pollution Control Investigation", J. Air Poll. Control Assoc., 10, 441, 1960.

35. "Standard Method for the Measurement of Odour in Atmospheres (Dilution Method), ASTM D-1391, 185, Amer. Soc. Test. Mat., Philadelphia, PA, 1959.

36. "Determination of odour and Test Thresholds by a Forced Choice Acending Concentration Series of Limits", ASTM E-679, Amer. Soc. Test. Mat., Philadelphia, PA, 1979.

37. "Standard Recommended Practice for Referencing Supra Threshold Odour Intensity, ASTM E-544, Amer. Soc. Test. Mat., Philadelphia, PA, USA.

38. Duffee, R. and O'Brien, M., Odour, Science and Engineering Inc., 57 Fishy St., Hartford, CT 06120, USA.

39. Selection and Training of Judges for Sensory Evaluation of the Character and Intensity of Diesel Exhaust Odours, US Public Health Service Publication No. 999 - AP - 32.

40. Matsushitia, H., Arito, J., Suzuki, Y. and Sodo, R., "Determination of Threshold Values for Olfactory Perception of Primary Odour Substances", Ind. Health, 5, 221, 1967.

41. Wilby, F., "Variation in Recognition Odour Thresholds of a Panel", J. Air Poll. Control Assoc., 19, 96, 1969.

42. Benforado, D., Rotella, W. and Horton, D., J. Air Poll. Control Assoc., 2, 19, 1969.

43. Prince, R. and Ince, H., J. Appl. Chem., 314, 1958.

44. Hillman, T. and Small, F., "Characterisation of the Odour Properties of 101 Petrachemicals using Sensory Methods, J. Air Poll. Control Assoc., 24, 979, 1974.

45. Leonardos, G., J. Air Poll. Control Assoc., 24 (10), 979, 1974.

46. Sensidyne Odour Monitor Application Sheet, Sensidyne Inc., Clear water, Florida, 1988.

47. Bailey, J. and Viney, N., "Analysis of Odour by Gas Chromatography and Allied Techniques", Stevenage, Warren Spring Laboratory, Report No. LR 298 (AP), 1979.

ENVIRONMENTAL MANAGEMENT AND LOSS MEASUREMENT IN DAIRY PROCESSING

J. Palmer, National Dairy Products Research Centre, Teagasc, Moorepark, Fermoy, Co. Cork, Ireland

ABSTRACT

The primary objectives in any waste management programme are to minimise product loss without affecting product quality and to dispose or treat the unavoidable losses in a safe and environmentally acceptable manner at an economical cost. Management of waste is an extension of good manufacturing practice and therefore should be included in the overall quality assurance plan. Reduction of losses firstly involves daily monitoring of all effluent for volume and strength and good quality control of product and byproducts. Regular data acquired from monitoring and quality control can be used to improve process efficiency where necessary. The data from daily records should also be used to set target figures for overall losses and losses occurring within specific processes. A good waste management plan will also serve as an aid in preventing overloading of the waste treatment plant and the resulting consequences of water pollution.

© 1990 IOP Publishing Ltd

INTRODUCTION

About four and a half billion litres of whole milk (1 billion gallons) are processed in the Irish Dairy Industry annually. It is estimated that at least 1% of milk processed is lost to waste representing a cost in the region of £10 x 10^6. Reducing this loss by as little as 10% would save £1 x 10^6 annually. Losses in dairy plants occur through the effluent drains, in product exceeding specification and in byproduct streams such as skim milk, whey and casein wash water. Another source of loss occurs in milk or casein drying due to escape of low density powder particles in the exhaust air. Reduction of losses firstly involves daily monitoring of the separate process waste streams and good quality control of products and byproducts. Figs. 1 and 2 summarize the salient points of loss management and loss measurement in the Dairy Industry. Flow measurement and flow proportional sampling are essential for monitoring losses. The data generated can be used to set target figures for losses occurring within specific processes.

FIG. 1. **ENVIRONMENTAL MANAGEMENT AND LOSS MEASUREMENT IN DAIRY PROCESSING.**

INDIVIDUAL DAIRY PROCESS WASTE STREAMS

FLOW MEASUREMENT AND SAMPLING

ANALYSIS

RESULTS - SET TARGETS

REDUCE LOSSES TO MEET TARGETS

FIG. 2. **LOSS MEASUREMENT TECHNIQUE (LMT).**

A. **TRADITIONAL METHOD:** MATERIAL BALANCE INACCURATE FOR DAILY USE DUE TO STOCK CARRYOVER AND ERROR IN MEASUREMENT.

B. **L.M.T.:** REQUIRES MEASURING ALL LOSS SOURCES IN EACH PROCESS.

C. **NECESSARY INPUTS:** FLOW MEASUREMENT AND SAMPLING EQUIPMENT ANALYSIS OF EFFLUENT, BY-PRODUCT AND PRODUCT

D. **EXPRESSION RESULTS:** AS KG OF PRODUCT LOST/DAY.

E. **SET TARGET FIGURES:** FOR LOSSES ON BASIS OF GOOD OPERATING PRACTICE.

F. **AVOIDABLE LOSS KG/DAY:** = TOTAL LOSS - TARGET VALUE.

Flow Measurement: Unfortunately there is no ready made system which gives accurate readings and which can be inserted in an effluent drain without doing some site work. Weirs and flumes in which the effluent level is measured are used in open channels. Portable weirs are useful for short term measurement of effluent flow before a permanent installation is made. Magnetic flow meters are accurate and can be used for effluent measurement provided that the pipe is always full. They require fitting a section of pipe of appropriate diameter to an inlet sump to maintain the pipe full of liquid and to have free discharge to an outlet chamber. (Fig. 3). All flow metering systems should be regularly calibrated for accuracy.

Sampling: A truly representative sample of effluent is difficult to obtain. Automatic samplers which are controlled by the flow metering equipment may be set to sample flow proportionately at a rate of 100ml per 1 - 2 m^3 of flow. The diameter of the sampler suction tube should be at least 1 cm and should be positioned in the effluent drain downstream from the flow measuring site and where maximum turbulence is achieved. The sampler should have a facility for backflushing the suction tube before each sampling and the sample container should have a volume capacity of about twice the daily sample volume to facilitate thorough mixing before subsampling, which is absolutely essential. When sampling effluents high in fat such as that from butter manufacture, the sampling container for holding the sample should be kept at 45°C. All sample containers except that used for treated effluent, should have sufficient hydrogen peroxide added as preservative to give 400 mg/l in sample collected (1.35 ml of 30% H_2O_2 per litre of sample). The sample container used for treated final effluent collection should be stored at 4°C.

Sampling of constant flow byproduct streams such as skim milk, whey, and casein wash water may be achieved by means of solenoid valves operated by a timer which are installed on the discharge line.

Analysis: The appropriate analysis for a particular waste or byproduct stream will vary depending on the process giving rise to the waste. In the case of effluents a measure of the organic content is essential for determining the organic loading for effluent treatment purposes. The chemical oxygen demand (COD) measures the oxygen consumed by the organic constituents of the waste on digestion with chromic acid and is the most common method used to measure the strength of untreated dairy waste. The relationship between biochemical oxygen demand (BOD) and COD falls within the range 60 to 70% of the COD for most untreated dairy wastes whereas the figure for fully treated or final effluent may be as low as 10%.

The COD test must be supplemented by other tests which measure fat, protein and lactose or equivalent in order to measure the amount of product or component lost to effluent. The tests for monitoring losses in effluents and byproducts arising from various dairy processes are given in Table 1.

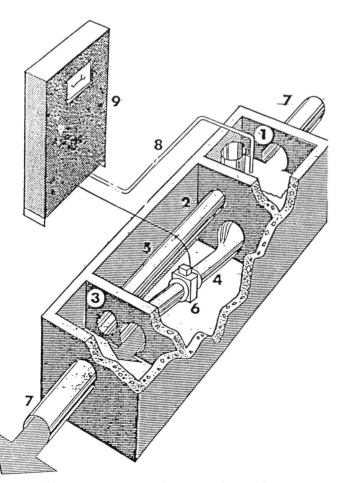

1. Inlet Manhole
2. Meter Chamber
3. Outlet Manhole (benched up)
4. Meter Pipework
5. Failsafe overflow pipe
6. Magnetic Flowmeter
7. Inlet and Outlet Sewer
8. Sampling tube
9. Weatherproof cabinet with sampler, converter, counter and recorder

Fig. 3: Magnetic Flow Metering Assembly.

TABLE 1: Appropriate tests for effluent and biproducts arising from various dairy processes.

Process	Effluent Tests	Byproduct Tests
Milk reception including pasteurisation cooling and storage	COD Fat	- -
Milk reception including separation, pasteurisation and cream storage	COD Fat	Fat in skim milk

Milk drying
  Skim milk
  Whole milk
	
	COD
	COD
	Fat
		
		-
		-

Butter	COD Fat	Fat in butter milk
Cheese	COD Fat Curd	Fat and curd in whey
Casein	COD Casein	Fat and casein in whey Casein in wash water

Expression of losses: As can be seen from Table 1, COD and fat describe effluent losses for several dairy processes namely milk reception & separation, milk drying and butter manufacture. Since the COD contributed by fat and solids non fat (SNF) approximate to 2.5 parts COD per part of fat and 1.1 parts COD per part SNF (Table 2) we can convert the COD value to product lost. The following examples illustrate this conversion of COD to product loss:

Effluent source -	Milk reception
Effluent volume -	200 m^3
Effluent analysis -	COD 2000 mg/l
	FAT 355 mg/l
COD due to fat =	355 x 2.5 = 888 mg/l
COD due to SNF =	2000 - 888 = 1112 mg/l
SNF =	1112/1.1 = 1011 mg/l
Kilogram of COD =	2000 x 200/1000 = 400
Kilogram of fat =	355 x 200/1000 = 71
Kilogram of SNF =	1011 x 200/1000 = 202

In the case of cheese effluent, the fat and curd values obtained can be converted to COD using factors of 2.5 and 1.9 respectively (Table 2) and the remainder of the COD expressed as SNF or as whey lost in m^3. Similarly for casein effluent the casein value is converted to COD by using a factor of 1.4 and the remaining COD expressed as whey lost.

TABLE 2: Chemical oxygen demand value of some milk products and constituents.

Product	Kg COD/m^3	Kg COD/Kg
Whole milk	180	0.174
Skim milk	100	0.097
Separated whey	63	0.061
Separated butter milk	113	0.109
Milk fat	-	2.5
Milk protein	-	1.45
Casein	-	1.36
Lactose	-	1.0
SNF in skim milk	-	1.1
SNF in whey	-	1.0

Target losses for effluent: Losses that occur during processing may arise due to, accidental spillage or leakage, desludging of separators or clarifiers and the cleaning of storage or balance tanks. Losses from these sources should be minimised. Recovery of product before cleaning should be maximized. When both of these are efficiently carried out, the measured losses are the target losses to be aimed for. The target loss figure for any process may be set by determining the losses over about 20 days after implementation of good management techniques for loss reduction. The loss figure in kilograms for any processing day minus the target loss figure may be taken as the avoidable loss for that day. Expressing loss values as % of raw material processed or as product manufactured or as fraction lost per unit processed can be misleading when applied at minimum or maximum throughput as losses are not directly related to throughput. Target loss figures for products depend on how precisely one can meet product specification without being under specification e.g. moisture in butter less than 15.7% could be taken as product loss. Product overweight also depends on what can be consistently achieved in terms of a specified weight without being under the specified weight.

Loss values which were obtained in several Irish factories are shown in Table 3.

TABLE 3: Loss values obtained for dairy effluent in various factories.

	% lost of total received as:		
	COD	Fat	SNF
Cheese	1.00 - 1.60	0.60 - 1.40	1.30 - 1.70
Milk intake & separation & pasteurisation	0.40 - 0.50	0.50 - 0.60	0.20 - 0.50
Milk intake & pasteurisation	0.75 - 1.14	1.00 - 1.42	0.50 - 0.67
Evap/Drier			
Skim milk	0.76	0.05	0.73
Whole milk	0.75 - 1.14	0.80 - 1.11	0.70 - 0.93
Butter	0.40	0.40	0.09
Casein:Acid	0.7	-	0.85
Casein:Rennet	1.6	-	0.84

Target loss figures of 750 mg fat in skim milk or cheese whey and 50 mg of curd or casein in cheese and casein whey are achievable. A casein target loss figure of 50 mg of wash water and 5000 mg of fat/l of buttermilk should be achievable.

In-line waste measurement: Significant product losses occur at the start-up and shut-down of equipment such as milk evaporators and heat exchangers. An evaporator economy device consisting of a multi-electrode conductivity probe, transmitter and microprocessor was used to detect the milk/water interface in an evaporator at start-up and shut-down. The microprocessor controlled the operation of valves on the lines to effluent, recycle or dryer (Fig. 4). An almost linear relationship was found between conductivity in microsiemens/cm and total solids (%) up to 25% TS. Conductivity levels of 2000 and 6500 microsiemens/cm corresponded with solids levels of 4 and 18% which were the target set points chosen for discharge to effluent or for recirculation to the drier. The use of the conductivity sensor reduced the losses to effluent per process run by an average value of 100 kg skim milk powder or 0.23% of skim milk processed.

Recording results of loss measurement: The losses occurring in any day which are over target should be recorded in a spread sheet setting out the amount of product manufactured, product lost to effluent, to byproduct, to over specification and to overweight. The financial cost of the product lost should also be recorded. The daily loss figure can be carried forward to give cumulative figures for specific periods.

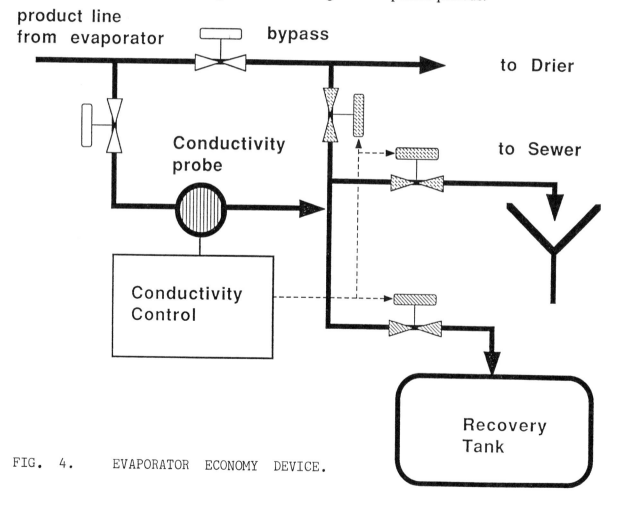

FIG. 4.   EVAPORATOR ECONOMY DEVICE.

AIR QUALITY MONITORING IN THE UK

J. Wilken, Warren Springs Laboratory, U.K.

ABSTRACT

This presentation describes national air quality monitoring programmes carried out on behalf of the UK Department of the Environment by Warren Spring Laboratory. These include networks for assessing acid deposition, ozone, oxides of nitrogen, smoke and SO_2, lead and other particulates. Taken together, these consititute the most comprehensive infrastructure for air quality monitoring throughout the United Kingdom.

Monitoring objectives and methodologies are discussed, with particular reference to modern instrumented networks utilising data telemetry. Some recent important findings from the rural ozone and urban nitrogen dioxide monitoring networks are also briefly considered.

© 1990 IOP Publishing Ltd

THE LEGAL AND POLICY FRAMEWORK FOR ENVIRONMENTAL MONITORING IN THE EUROPEAN COMMUNITY

Mr. L. Cashman, Directorate General Environment Nuclear Safety and Civil Protection, Commission of the European Communities, Brussels, Belgium

ABSTRACT

The purpose of this presentation is to set the technical aspects of environmental monitoring in a broad E.E.C. legal and policy context. Environmental rules and policies adopted by the E.E.C. have already brought about considerable changes in the environmental law of Member States, and will continue to do so in the future. Many Community rules and policies entail the creation of monitoring regimes, and more make such regimes either advisable or desirable. The presentation will attempt to :

(a) briefly show why the E.E.C. has developed a body of environmental rules and policies ;

(b) explain something of how E.E.C. rules and policies are developed and standards set, making reference to the importance of key texts such as the Single European Act and the four environmental action programmes, and indicating how directives evolve from conception to implementation ;

(c) relate present and likely future rules and policies to a hypothetical pharmaceutical project, with particular reference to monitoring, mentioning :

- the possible future importance of strategic environmental assessment,
- the existing importance of project impact assessment,
- the need to take account of numerous Community authorisation requirements,
- requirements and opportunities in regard to clean technologies,
- the possible relevance of ecological labelling and a proposal on civil liability for damage caused by waste,
- the interconnected monitoring roles of national authorities (including environmental agencies), local authorities and industry,
- the rights of citizens in regard to monitoring data,
- the role of the Commission and the European Environmental Agency.

© 1990 IOP Publishing Ltd

1. Introduction

Over the last 15 years, European Community environmental law has fundamentally shaped Irish evnironmental law. In many key aspects, European Community environmental law is Irish environmental law.

The sweep of the Community's legislative role can be gleaned from a glance at recent proposals, which have ranged in subject-matter from habitat protection to municipal waste incinerators.

The fertility of the Community in environmental legislation has considerable implications for those concerned with environmental monitoring. These implications will be examined in the paper, with particular reference to the pharmaceutical industry.

For the purposes of this paper, two broad categories of environmental monitoring can be distinguished:

(a) <u>Monitoring for compliance with particular rules or standards (compliance monitoring)</u>

Such monitoring may range from multi-media monitoring of particular point sources of pollution to general monitoring of air quality, or water quality for a particular classification of water (e.g. a bathing water). Monitoring requirements may be explicit or implicit. For example, if it is required that discharges of particular dangerous substances be authorised so as to help attain certain quality objectives, than it is reasonable to infer that there should be post-authorisation monitoring to make sure that objectives are actually achieved.

(b) <u>Optional monitoring</u>

Sometimes environmental monitoring may not be strictly necessary to control compliance with particular rules or standards, but may nevertheless be useful or advisable as a management tool.

European Community environmental law is of significance to both types of monitoring. However, before I discuss this significance, I would like to say something about the reasons which lie behind the creation of a Community environmental policy in the first place, and how this policy results in the creation of rules and standards in practice.

2. Why a European Community Environmental Policy ?

The European Economic Community is a supra-national Community founded on the Treaty of Rome which dates from 1957. The Treaty defines the activities of the Community and sets out the powers of its four institutions, the Commission, the Council of Ministers, the Parliament and the Court of Justice.

The original Treaty made no explicit mention of the environment, but by the 1970's it was felt necessary for the Community to develop environmental rules and policies. The Community is chiefly but not exclusively an engine of economic growth, having as major goals the constant improvement of living and working conditions and the harmonious development of Member State's economies.

The Paris summit of October 1972 marks a major turning point. In dedicating themselves to the development of a Community environmental policy, the then Community heads of state declared that economic expansion was not an end in itself : its first aim should be to enable disparities in living conditions to be reduced, and it should result in an improvement in the quality of life as well as in standards of living.

The notion that the EEC is about quality of life (amongst other things) is fundamental to an understanding of why the Community has developed an environmental policy.

However, there is another crucial reason: a true common market aims at the creation of a level playing field for industrial competitors. Environmental controls may affect competitive conditions in at least two ways - at the manufacturing stage as a factor of production ; and at the marketing stage as potential barriers to trade.

While quality of life may represent the higher aim of Community environmental policy, the need to avoid unequal conditions of competition is invoked over and over again to justify particular pieces of environmental legislation.

So, if you like, there is a hard edge and soft centre to Community environmental policy.

3. How is Community Environmental Policy Developed ?

Community environmental policy has come a long way since 1972. Of great significance is the fact that the Treaty of Rome has now been amended by the Single European Act so as to incorporate explicit environmental provisions. These are to be found in Articles 130r, s, t and 100 A of the amended treaty.

Community action on the environment is given three fundamental objectives :

(i) to preserve, protect and improve the quality of the environment ;
(ii) to contribute towards protecting human health,
(iii) to ensure a prudent and rational utilisation of natural resources.

In other words, Community action is not confined to measures aimed at environmental protection in the strict sense, but extends to health and safety measures and measures intended to conserve natural resources. Examples of such measures are to be found in the so-called Seveso directives and the drinking water directive.

Apart from setting objectives, the amended Treaty espouses the principles of preventive action, rectification at source and polluter pays.

Also worth mentioning is the possibility (under Article 100A) of legislating on the environment by means of qualified majority voting.

The Single European Act represents a landmark in the development of Community environmental policy, but before its adoption the Community had already endorsed a series of wide-ranging environment action programmes. The first dates from 1973, and the latest (the 4th) will extend until 1992. In some detail, these define, sector by sector, the principles of Community action and the areas in which initiatives are to be taken.

4. How are Specific Initiatives Developed ?

Firstly, due account must be taken of the express constitutional foundation now provided by the amended Treaty. Secondly, initiatives are very much guided by the broad sectorial policy aims set out in the 4 action programmes, which tell us what we can "expect" from the Community during their currency.

The formal proposal of initiatives lies within the exclusive competence of one institution, the European Commission. In practice, an intention to take an initiative will usually be signalled in advance in an action programme.

The work of preparing a proposal will usually begin with the commissioning of a detailed study from a specialised institute or consultancy. If, for example, the Community proposes to regulate emission limit values for discharges of certain dangerous substances, this would probably take the form of an ecotoxicological study on the substances which might cover existing regulatory regimes both within and outside the Community, analysis methods and so forth.

The study will then be submitted to national experts from each Member State for their comments and observations.

Ideas for a text having thereby been gathered, drafting will begin. This is the task of the relevant technical service of Directorate General XI, Environment Nuclear Safety and Civil Protection (one of 23 directorate-generals or departments within the Commission). The first draft will usually form the subject of fresh consultations with national experts, and then the text of the formal Commission proposal will be prepared.

Before adoption by the Commission, it will be circulated for comments and observations to all the other services concerned within the Commission.

The process of consultation continues after adoption of a text by the commission. The Economic and Social Committee (ECOSOC) and the European Parliament are both formally invited to consider the proposal.

In practice, ECOSOC exercises a formal rather than an active role. Not so the European Parliament which, through its Environment Committee is increasingly submitting pertinent amendments to the proposals which come before it.

After all these consultations, the proposal must be agreed or rejected by the Council of Ministers. Only with the accord of the Council of Ministers can a proposal become law. In practice, a text will often be altered at this stage by a working group comprising representatives of the Member States and the Commission. The text finally adopted is very often a compromise text.

5. What Happens after Adoption of a Legal Text ?

Firstly, it should be said that, in environmental matters, the European Community usually legislates by means of directives. A directive is a legal instrument which is binding as to the results to be achieved but which leaves to Member States the choice of form or methods for achieving these results. In practice, it is generally necessary for a Member State to adopt national legislation transposing the provisions of a directive by a particular deadline. Failure to do so can result in all sorts of legal complications. It is essential to note that, once adopted, Community environmental directives create a bundle of rights and obligations, with consequences for national authorities, local authorities, individuals and industry.

Before passing on to a brief examination of these consequences in regard to the pharmaceutical industry, I would like to mention that many directives take cognisance of the fact that scientific advances often make revision of technical provisions desirable, for example as regards monitoring methods.

This is why directives frequently provide for the creation of Committees to assist in the adaptation of technical annexes to scientific and technical progress.

6. Monitoring issues for the Pharmaceutical Industry

It is not possible here to do justice to all the monitoring implications of Community environmental law for the pharmaceutical industry, but I would like to try to relate some of these implications to various crucial decision-making stages in the development of the industry as well as to plant operation.

7. Strategic environmental assessment

Environmental impact assessment (E.I.A.) for individual projects is now very much a reality for the pharmaceutical industry in Ireland. However, the development of a pharmaceutical industry is not simply a matter of accumulating individual projects. National industrial policy and strategic land-use planning are of key importance.

When the Commission was first exploring the possibility of a Community system of environmental impact assessment, it was its intention to extend such a system to plans, policies and programmes. In the event, the so-called E.I.A. directive adopted in 1985 confined itself to projects.

However, environmental assessment of plans, policies and programmes (or strategic environmental assessment) is by no means a dead concept.

The fourth Environmental Action Programme commits the Commission to considering it, and it is not unlikely that a proposal will be forthcoming during the next few years.

S.E.A. might perhaps affect the pharmaceutical industry in two ways. Government strategies for developing or regulating the industry might be formally required to explore the general environmental implications of the industry, and to consult the public on them.

The earmarking of particular locations for chemical and pharmaceutical industries might also have been subjected to environmental assessment - even before any project is proposed. Again, the public would have to be consulted.

The object of S.E.A. is not to impose further bureaucratic impediments to development, but rather to integrate environmental concerns into higher tiers of decision-making. It is an example of the principle of prevention espoused in the amended Treaty.

In terms of its implications for monitoring, S.E.A. will undoubtedly increase the need for general and particular environmental data. Since it may be desirable or necessary to separate decision-making and assessment functions, the role of environmental agencies (such as that envisaged by the present Irish government) will probably be an important one.

8. Project Impact Assessment

Whereas S.E.A. is something for the future, Project Impact Assessment is very much of the present.

I believe the issues raised by the impact assessment directive have by no means settled, and that we will see many developments in terms of the legal interpretation of particular provisions, as well as in terms of the practice of making assessment.

The impact assessment procedure turns on the information to be provided by the project proponent, and this may vary according to particular circumstances. However, as a minimum, Article 5.2 of the directive provides that the information shall include :

- a description of the project comprising information on the site, design and size of the project,
- a description of the measures envisaged in order to avoid, reduce and, if possible, remedy significant adverse effects,
- the data required to identify and assess the main effects which the project is likely to have on the environment,
- a non-technical summary of the information mentioned in indents 1 to 3.

So far as monitoring is concerned, I believe a particular importance must be attached to the words "the data required to identify and assess the main effects which the project is likely to have on the environment". To understand what is meant by "environment" we must look to Article 3 which speaks of the following factors :

- human beings, fauna and flora,
- soil, water, air, climate and the landscape,
- the inter-action between the factors mentioned in the first and second indents,
- material assets and the cultural heritage.

For something as major as a pharmaceutical project, information which does not include full monitoring results for baseline conditions may be open to challenge on the basis that it fails to provide the data for assessment of the main effects on these factors. Such monitoring would be the task of the proponent.

A criticism that may be levelled at the present directive is that it does not make provision for an environmental audit after a project has actually been completed. Without auditing, project E.I.A. is an incomplete methodology.

However, even in the absence of a specific legal requirement, there are good reasons for carrying out monitoring of the actual as distinct from the predicted effects of a project on the environment. Developers should not look on E.I.A. as simply an exercise to satisfy planners and the public. It should be seen as a tool of environmental management. In the context of increasingly stiff environmental regulations, good environmental management will pay economic dividends, and there is a clear interest for developers in making sure that those they pay to make environmental predictions for them should actually get it right. It is interesting to note that major management consultancies have begun to include environmental auditing in their portfolio of services (auditing, in this context, seving a wider purpose than simply following up E.I.A.'s).

9. Authorisations

Apart from project E.I.A., regard should be had to the range of authorisaiton requirements which European Community environmental law has created in regard to water discharges, air emissions and waste disposal. With these, in some cases, come explicit monitoring requirements. Even where there is not explicit provision, monitoring may sometimes be considered an implicit requirement. Of course, in this context, by monitoring I mean compliance monitoring, a function of public authorities.

It is not possible to discuss all these authorisation requirements in detail here, but one or two are perhaps worth mentioning.

For example, Directive 76/464/EEC on pollution caused by certain dangerous substances discharged into the aquatic environment of the Community provides for a potentially far-reaching pollution reduction system for a wide range of dangerous substances objectives, authorisations and emission standards, with post-authorisation compliance monitoring probably an implicit requirement. To date, Article 7 has been largely ignored, though the Commission is now making efforts to secure improved compliance.

As regards air emissions, Directive 84/360/EEC on the combating of air pollution from industrial plants is of particular note. This directive subjects to prior authorisation the operation of a significant range of industrial plants (including, for example, the toxic waste incinerators commonly associated with pharmaceutical projects) and provides for the compliance monitoring of emissions.

10. Other Requirements

Apart from authorisations, other requirements may have relevance for monitoring, for example, the so-called Seveso directives (82/501/EEC and 87/216/EEC) on the major accident hazards of certain industrial activities. These directives apply to the pharmaceutical industry, amongst others, and to existing as well as new industrial activities. Their purpose is, in the first place, to try to prevent major accidents, and, in the second place, to minimise their effects.

An onus is placed on manufacturers to identify major-accident hazards, and adopt appropriate safety measures. Proper monitoring of plant operation can be considered essential under this directive.

I have already mentioned Directive 84/360/EEC in regard to authorisations. It has a further significance in as much as it makes provision for the application of best available technology to new plants as well as existing ones. Together with the Member States, the Commission is working on the preparation of technical notes to define, sector by sector, what constitutes best available technology. A number of technical notes have already been prepared, including one on the incineration of hazardous wastes.

It may be worth mentioning here that the Commission is giving serious consideration to adopting a multi-media sectorial approach to emissions and discharges. This might mean that, in the future, at Community level there would be a comprehensive set of rules and standards prepared specifically for the pharmaceutical industry. This would be in place of the present rather piece meal approach.

11. Clean Technologies

Two topics have been very much exercising Community environmental policy-makers recently - clean technologies, and fiscal incentives, both to some extent inter-related.

I can claim no expertise on clean technologies, but it seems to me that a clean technology should not only be a technology which minimises adverse environmental effects : it should also be a technology which is capable of monitoring and assessing those effects. In other words, it should provide for accurate monitoring.

The Community is encouraging clean technologies in two ways :

- by regulatory measures (such as Directive 84/360/EEC already mentioned) ;
- by supportive measures, such as the ACE programme and NETT.

At present, the Community is exploring ways of encouraging clean technologies (amongst other things) through fiscal incentives.

An impetus to the introduction of clean technologies (and better environmental monitoring) may result from the adoption of a current Commission proposal on civil liability for damage caused by waste. This proposal would introduce a principle of strict liability on a producer for damage caused by waste, damage not being confined to classic damage in the sense of personal injuries and damage to property, but also extending to environmental damage. Because it will be easier to sue for such damage, there will be an incentive to closely monitor and control industrial processes so as to minimise the production of waste.

The possible establishment of a Community system of ecological labelling (on which work has begin within the Commission) may serve as another incentive for the adoption of clean technologies, since it would probably endorse a "cradle-to-grave" approach, covering the manufacturing as well as the consumption and post-consumption stages of a product's life. A Community ecological label should prove a valuable marketing point for those companies which manage to fulfill what will probably be strict pre-conditions for acceptance.

12. Monitoring Roles

The foregoing has given some indication of the relevance of Community environmental policy and law for compliance and optional environmental monitoring.

Compliance monitoring is for the most part the task of public authorities. The tendency is towards more extensive and more sophisticated monitoring and this may be an issue for certain under-resourced local authorities, and may be one justification for giving a monitoring role to a body with a centralised pool of expertise, such as a national environmental agency.

Although compliance monitoring is essentially a public function, it should be borne in mind that some degree of self-monitoring is required under Community directives, for example the Seveso directives and the industrial plants directive already mentioned earlier.

Even where there is no requirement to monitor, optional monitoring may be a good idea for industry both to avoid the consequences of non-compliance with particular standards, and to gain the advantages that accrue from good environmental practice.

13. Rights of Citizens

Community environmental policy and law is in a state of rapid evolution, with new concepts, approaches and directions becoming apparent all the time. However, for some time it has been clear that there is a steady trend towards empowering citizens and environmental groups to act in environmental matters.

One good example is the public consultation provided for in the environmental impact assessment procedure. Another is the facilitation of legal proceedings for environmental damage under the Civil liability proposal. Yet another example is to be found in a proposal for liberalising public access to official environmental data (largely agreed by the Council of Ministers in the Environmental Council of March 1990).

These developments, combined with increasing public vigilance in environmental matters, make good environmental practice by industry an imperative.

Even if it can be guaranteed that certain public authorities will take a lax approach to polluters, it cannot be guaranteed that the public will do likewise. Easier access to environmental data will give more teeth to environmental groups.

14. The Role of the Commission and the European Environmental Agency

As for a strategy of shifting location within the Community to take advantage of more lax regulation conditions, there may be some short term advantages, but I would suggest that, in the medium and longer-term, such a strategy would be mistaken.

For one thing, the mobility of multinational industry is being increasingly matched by the networking of environmental pressure groups. There is increasing evidence that many of those are seeing their role in pan - Community terms, and they are probably set to make increasing use of the rights and obligations that flow from Community environmental law.

Some of this evidence can be found in the large increase in environmental complains made to the Commission over the last three years.

Apart from preparing directives, the Commission has the role under the EEC Treaty of making sure provisions of Community law are respected. The operation of a complaint procedure is an aspect of this role.

The complaints procedure works as follows. If the Commission receives a communication (normally a letter, but sometimes video recordings are sent) alleging facts which would constitute a breach of some provision of Community law, a complaint will be registered and an investigation camed out. If the investigation finds substance in the complaint, legal proceedings can be commenced against the Member State.

In this way unsuspecting industrial plants can find themselves under scrutiny by the Commission anywhere in the Community. Of course, it will be national authorities who will be held accountable for any failures of implementation, but this is bound to carry indirect consequences for plants breaking the law.

In carrying out an investigation, the Commission will often look for evidence that all required monitoring has been carried out.

Apart from the complaints procedure, it is worth mentioning that the role of controlling compliance is also partly fulfilled through the reporting systems which some directives institute. These provide for the submission of national reports on implementation at regular intervals from the Member States to the Commission. These reports sometimes take the form of syntheses of monitoring data collected in a Member State.

In conclusion, I would like to briefly mention the European Environment Agency and the European Environment Information and Observation Network. In January 1989, Commission President Jacques Delors proposed the creation of a Community environmental agency in a speech to the European Parliament. The Community quickly and favourably responded, and in a little over a year (at the March 1990 Environmental Council) Member States largely agreed a formula.

The structure of the Agency and the Information and Observation network it will operate is deliberately intended to be loose and decentralised, with national focal points and a number of Community thematic centres.

A two-stage development of the Agency's role is foreseen. Initially, it will act as an information gatherer and co-ordinator to assist in the implementation of Community environmental policy. It should therefore play an important role in enabling good use to be made by the Community of monitoring data collected at national level.

At a later stage, it has been agreed that proposals will be presented for a possible extension fot he Agency's functions. In particular, the Agency may become associated in the monitoring of implementation of Community legislation.

Whether this will ultimately result in a Community environmental inspectorate remains to be seen.

NETT: A KEY FOR INFORMATION ON ENVIRONMENTAL TECHNOLOGY

Marie Claire NAS, NETT, Brussels, Belgium

ABSTRACT

3.228 billion ECU : this represents the world market on environmental protection of which 40% is accounted for by the European market. Today, a technological answer has to be found to environmental problems. Action is in the hands of science and industry that together bring about technological solutions that respond to the motor of action in the "hot spot" wave of this decade : legislation responds to a call for a cleaner environment, where public opinion also plays more and more an important role. Environment has become an integral part of all other Community policies : a guide of conduct. The control on these stricter norms will for a great part be in the hands of the newly created European Environmental Agency. Through stricter control on complying with highier standards industry will be forced to participate in the application of clean and low-waste technologies. Here, for example, is a vast occasion for the pollution control equipment industry - and in particular for the sensor production industry - to expand their markets.

Today, environmental protection has become the new challenge of the 90"s. This was already foreseen by the CEC, that has launched an European centre for information on environmental matters. The CEC realized that obtaining correct information, is essential for solving environmental problems. The ncessary information on environmental monitoring and control is crucial to see how an industry, public or private, is in compliance with legislation.

NETT, the European centre for information on environmental matters, is an independent organization, business-minded, located in Brussels, and provides to its members the following services :

- specialized on-line database service,
- personalized service suited to specific request,
- "meeting" service.

© 1990 IOP Publishing Ltd

I. **INTRODUCTION**

75 billion ECU, or 57 billion Irish Pounds. Voilà, this represents the world market on environmental protection, 40 % of which represents the European market. And the growth of the market is tremendous : 6 to 10 % annually. Environment is becoming business, big business.

Today, a technological answer has to be found to an environmental problem. What developments in environmental technology and in environmental monitoring do we need to envisage in the next decade, up to the year 2.000. That is the central question.

If philosophy has been the motor of consciousness of environmental problems - and it should continue to fuel the action, it should continue to support, through its pertinence, public awareness. Now, today, action is in the hands of science and industry that, joined together, should think about, develop and finalize, technological solutions. Environmental problems are many, EEC transboundary ones, national, regional and local. If philosophy has been the motor of consciousness, action is to be found in technology. The motor of action in environmental matters is legislation.

II. MONITORING : THE EUROPEAN ENVIRONMENT AGENCY

European legislation has become stricter, imposing more constraints on the industry. Since the Single European Act, environmental protection policy has become an integral part of all other community policies, and when harmonized, domestic environmental regulations will become applicable. They are the known references, the basis on which environmental monitoring is established. Monitoring is the framework linking legislation to industrial reality.

Recently, the Commission has approved the creation of the European Environment Agency. What is it going to do ? First of all, it is going to measure the state of the environment. The waste water industry for example, is going to be confronted with a high quantity of measurement equipment. They are put at regular intervals along water lines carrying waste water and measure at different stages of the industrial processes, and at different time intervals, the level of pollution. A huge expansion of the sensor industry is expected.

Secondly, it will be monitoring. The Agency's role will be crucial in comparing consequently environmental pollution reality with the established references : the law.

A problem is then created when reality and law are not in line.

It can readily be established that solving problems in environmental areas depend on the decision to invest in new, clean, and low-waste technologies.

III. MONITORING : THE INDUSTRY

What does this mean at the level of a plant ? Being invaded by measurement equipment, a continous verification process will be going on.

Do we comply with applicable domestic legislation ? Environmental impact assessment becomes a crucial aspect of long and middle term strategy.

Then, action has to be taken. Investing in new technologies demands either re-assessment of your manufacturing processes, or raw materials to be recycled. A choice is to be made. Another basis has to be found on which your competitiveness can be established.

The necessary information for the public or private decision-maker, has to be readily accessible, and relevant, taking into account the increasing complexity of its legal and technological aspects.

Before investing, questions are raised :
- do I invest in the best technology ?
  * did I make a right choice in recycling or not ?
  * how do I get to know about new technologies ?
- do I invest in the best place ?
  * shall I act in partnership ?
- do I take into account relevant legislation ?
  * when is a specific EEC directive becoming domestic law ? (85/337/EEC - 27 June 1986 - on environmental impact assessment)
- do I invest at a reasonable cost ?

* economic aspects : what about profitability of my choice ?
* what are the limits of my human and financial resources ?
* financial aspects : can I get financial support, from the EEC, or are there other sources; they also determine the speed at which action can be taken.
- finally, should possibilities of merger or acquisition be considered ?

These questions are raised before investing, but afterwards the same questions are applied to check retro-actively, whether the right choice has been made. Because then, the whole process of monitoring starts again.

The Commission has understood the danger of not centralizing information, not being complete or correct. And above all, not being adapted to an evolution in a fast growing sector that is the biggest market in the last decade of this century.

NETT has been created to respond to these questions.

IV. NETT : NETWORK FOR ENVIRONMENTAL TECHNOLOGY TRANSFER

NETT has been set up by the EEC as the European information centre for environmental matters.

It is a non-profit international organization established with the strong support of the Commission.

It is organized around three services :

1. specialized on-line database services,
2. a personalized service suited to specific requests,
3. a "meeting" service.

1. <u>Specialized on-line database services</u>

The specialized on-line database system : DATANETT.

Through a modem linked to your PC, access will be obtained to DATANETT : one gateway that leads into four main directions.

First of all, there are the internal databases : NETTBASE, the on-line directory of members in which extensive profiles are given, TECHNETT an index in which members describe their activities, the fields of their specialization, their technological processes and in which offers and demands on technology are listed. Also an on-line version of the organization's newsletter : NETTINFO.

Secondly, the direction in which specific databases on environmental issues are assembled. They are divided in two parts : the first contains specific environmental databases on water, air, soil, noise, management.

The second part is concentrated on EEC and national environmental legislation, EEC programs and call for papers, as well as all data on the impact of Europe 1992.

The third direction lists over more than 1.000 different databases on areas linked to environment, such as chemistry, toxicology, physics, engineering, agriculture, business and finance, law, etc...

A special user service is added : SMARTSCAN a tool that allows the user to access directly, with the help of keys, the specific databases you need to consult.

And finally, the fourth direction, facilitating communication through electronic mailing.

The advantages of such a gateway are evident :

- one single access language,
- one password,
- one single contract,
- one invoice,
- on-line communication between members.

2. <u>Personalized service suited to specific requests</u>

Through a simple phone call or fax specific requests can be submitted, like search for partners to obtain EEC financial aid or for the sale of licences for a newly introduced product.

3. <u>The "meeting" service</u>

Through this service, conferences, seminars, workshops and participation to fairs are organized. And set up as a "Club", members receive priviliged access to professional and social activities.

The important developments in environmental legislation have created a necessity for companies to adapt and to invest, for which the choice of the right technology is a must.

The quality of the information is vital for a cost-effective investment.

Information is change, continual change. Lesgislation is moving, re-evaluation of the context is important.

The European industry must take up this challenge. Now a European information centre for environmental matters exists : its name is NETT.

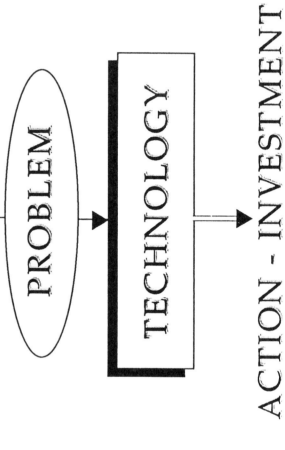